Current Human Cell Research and Applications

Series Editors

Nariyoshi Shinomiya, Department of Integrative Physiology and Bio-Nano Medicine
National Defense Medical College
Tokorozawa, Japan

Hiroaki Kataoka, Department of Pathology
University of Miyazaki
Miyazaki, Japan

Yutaka Shimada, Department of Nanobio Drug Discovery
Kyoto University
Sakyo-ku, Japan

This series covers basic and clinical research on human cells, including molecular diagnostics/targeted therapy, cell therapy, cancer stem cells, regenerative medicine, etc., and provides an up-to-date review of human-cell research. All volumes are contributed by leading experts in the field, and offer valuable resources for both cell biologists and clinical researchers in the areas of oncology, stem cell biology, regenerative medicine, and clinical medicine including gynecology, gastroenterology, etc.

Current Human Cell Research and Applications will be published in partnership with the Japan Human Cell Society

Nariyoshi Shinomiya • Yutaka Shimada
Editors

Applications of Cell Culture

Springer

Editors
Nariyoshi Shinomiya
Research Center for Biosafety, Laboratory Animal and Pathogen Bank
National Institute of Infectious Diseases, Japan Institute for Health Security
Shinjuku-ku, Tokyo, Japan

Yutaka Shimada
Department of Nanobio Drug Discovery, Graduate School of Pharmaceutical Sciences
Kyoto University
Sakyo-ku, Kyoto, Japan

ISSN 2522-073X ISSN 2522-0748 (electronic)
Current Human Cell Research and Applications
ISBN 978-981-95-0586-9 ISBN 978-981-95-0587-6 (eBook)
https://doi.org/10.1007/978-981-95-0587-6

This Springer imprint is published by the registered company Springer Nature Singapore Pte Ltd.
The registered company address is: 152 Beach Road, #21-01/04 Gateway East, Singapore 189721, Singapore

Preface

Although attempts at tissue and cell culture started over 100 years ago, we had to wait for the establishment of L cells in the 1940s for their practical use in experimental research. Since then, a variety of cells have been isolated from animal and human specimens, and many of them have been established as registered strains. Today, experiments using cultured cells occupy a key position in life sciences research, which provide objective evidence of biological activity. The information about the function of proteins, nucleic acids, and enzymes, the mechanisms of metabolism and intracellular signaling, the relationship between the cells, the pathways of carcinogenesis, degeneration, and aging, key players of differentiation and proliferation, and key signals of ontogenesis, all of which can be obtained from various cell culture systems. The Japan Human Cell Society has been involved in the establishment of numerous cell lines for over 40 years, starting with the cultivation of gynecological cancer cells and thereafter gastrointestinal cancers and hematological malignancies, based on the importance of the use of cultured cells. We have also disseminated various research results using those established cell lines to the world.

The publication of this book was planned to provide information from multiple perspectives on the types of research in which various experimental systems are introduced using cell cultures. Chapter 1 looks back on the history of the establishment of cell lines and outlines the research fields that cultured cells have been involved. Chapter 2 describes how dental pulp-derived stem cells (DSCs) are utilized as an example of primary culture, highlighting important perspectives in establishing clinical-grade standards for regenerative therapy and personalized medicine. Chapter 3 discusses the maintenance of established cells, appropriate subculture, and methods for inducing differentiation, providing practical guidelines for these processes. It focuses on three core factors: substrate coating, culture media composition, and environmental consistency. Chapter 4 focuses on the latest findings for the culture methods of pulmonary epithelial cells, particularly alveolar epithelial cells, and discusses the importance of cell culture in the development of effective treatments for intractable lung diseases. Chapter 5 takes a slightly different perspective and delves into the issue of cell line contamination and mix-ups that can distort research results, providing information on the importance of cell authentication and how to confirm the authenticity of cell lines. Chapter 6 focuses on the recent advancement in the cell culture of circulating tumor cells (CTCs). There are some limitations and an essential crux. The next challenge in this field is the establishment of easy, rapid, and inexpensive methods to cultivate CTCs recovered from individual patients. Chapter 7 provides an overview of organoid culture and its applications in cancer research by focusing on the current status of organoid-based research on endometrial and cervical cancers. It also describes the challenges and perspectives associated with organoid culture techniques in basic and translational cancer research. Chapter 8 focuses on limitations of long-established cell lines suffering from genomic drift, lack of clinical annotation, and underrepresentation of rare cancer subtypes—factors that compromise reproducibility and translational utility. It addresses challenges for dealing with them and encourages efforts to reinforce the role of cell lines as foundational tools in data-driven cancer biology for precision therapeutics. Chapter 9 describes the critical role of cancer cell lines in advancing rare-cancer research, focusing on their utility as scalable, experimentally tractable platforms for molecular and pharmacological studies. It emphasizes the importance of the culture of transparency and accessibility, which accelerates progress in the study of rare malignancies. Chapter 10 introduces research examples using induced pluripotent stem cell (iPSC) data and artificial intelligence (AI) in the field of neurodegenerative disease, especially focusing on the point that patient iPSCs generated by using reprogramming technology have the identical genetic background and can be differentiated into any cell types to construct a disease cell model.

Understanding the key concept of cell culture methods and their relationship to the development of diagnostic methods and therapeutics based on the current state of cell research can serve as a source of inspiration for new research and development tools. For research scientists using cell culture systems, this book provides deep insight into each research topic. It also helps recognize the uniqueness and novelty of cutting-edge science from a new perspective.

Shinjuku-ku, Tokyo, Japan Nariyoshi Shinomiya
Sakyo-ku, Kyoto, Japan Yutaka Shimada

Contents

About the Editors

Nariyoshi Shinomiya, M.D., Ph.D. His areas of expertise are microbiology/immunology and molecular oncology. His research interests include the radiosensitivity and apoptosis of tumor cells, Met tyrosine kinase function, and regulation of tumor growth, among others. He served as President of the Japan Human Cell Society from 2015 to 2023.

Yutaka Shimada, M.D., Ph.D., FACS He established more than 60 human esophageal cancer cell lines (KYSE-series) and 10 pancreatic cancer cell lines (KMP-series and TYPK-series). He also developed several gallbladder and bile duct cancer cell lines. His recent research focuses on establishing a standard method for culturing circulating tumor cells. He served as Vice President of the Japan Human Cell Society from 2015 to 2023.

Chapter 1
History of Cell Line Establishment and Application

Nariyoshi Shinomiya

Abstract Cell culture technology is not only a fundamental research tool for elucidating the functions of cells, organs, and living organisms but also brings immeasurable benefits to our society. Cell culture is applied in a wide range of fields, such as evaluating the efficacy and toxicity of drugs, developing and manufacturing vaccines and biopharmaceuticals, and applying to gene therapy, assisted reproductive medicine, and regenerative medicine, and it is no exaggeration to say that it plays a major role in the biotechnology industry and medicine. In this chapter, we will look back on the history of cell culture, which is the premise for reading this book, and summarize the important elements of cell culture, the basic considerations for advancing research and development, and its range of applications and future potential and also present the problems that cell culture faces. We also hope to use this to understand the current state of cell culture technology applications in advanced life science research, which will be shown in Chap. 2 and onwards.

Keywords HeLa · Cell culture · Immortalization · Authentication · Ethical issues

Abbreviations

ABC	ATP-binding cassette
AI	Artificial intelligence
CCLE	Cancer Cell Line Encyclopedia
CGAP	Cancer Genome Anatomy Project
CHO	Chinese hamster ovary
ES cells	Embryonic stem cells
HEK	Human embryonic kidney

N. Shinomiya (✉)
Research Center for Biosafety, Laboratory Animal and Pathogen Bank, National Institute of Infectious Diseases, Japan Institute for Health Security, Shinjuku-ku, Tokyo, Japan
e-mail: shinomi@niid.go.jp

N. Shinomiya, Y. Shimada (eds.), *Applications of Cell Culture*, Current Human Cell Research and Applications, https://doi.org/10.1007/978-981-95-0587-6_1

HGE Registry	Human Genome Editing Registry
HPV	Human papillomavirus
ICLAC	International Cell Line Authentication Committee
ICT	Information and communication technology
iPS cells	Induced pluripotent stem cells
IVG	In vitro gametogenesis
lncRNAs	Long noncoding RNAs
MDCK	Madin-Darby canine kidney
MDR	Multidrug resistance
miRNAs	Micro RNAs
MSCs	Mesenchymal stem cells
NCI	National Cancer Institute
NCI60	NCI's 60 Human Tumor Cell Line Anticancer Drug Screening
NIH	National Institute of Health
Rb	Retinoblastoma
SV40	Simian virus 40
TCGA	The Cancer Genome Atlas
UKCCCR	United Kingdom Coordinating Committee on Cancer Research
WHO	World Health Organization

1.1 The History of Cell Culture

1.1.1 HeLa Cells

The earliest established human cell line in the history of research is the well-known HeLa cell line, an immortalized cell line established from the cancer tissue of Henrietta Lacks, a female patient with cervical cancer. HeLa cells were first isolated by George Otto Gay, Margaret Gay, and Mary Kubicek; developed at Johns Hopkins Hospital in Baltimore, Maryland, in 1951 [1]; and reported in a paper the following year [2]. Since then, this cell line has been used in research for so long, largely due to the advancement of various cell culture techniques and the expansion and widening of the range of applied research, but above all the fact that the cells are immortalized and can be easily subcultured.

Immortalization of cells is the ability to proliferate indefinitely and escape from cell death. HeLa cells are infected with human papillomavirus type 18 (HPV18), which is involved in the synthesis of a protein that degrades the p53 tumor suppressor protein [3], and this is related to the immortal proliferation of HeLa cells. In addition, the HPV-derived E1 protein is thought to suppress the expression of genes involved in the cellular antiviral immune response, thereby giving a favorable condition for viral infection [4]. In other words, such immortalized cells have the property of generating their proliferation signals within the cells and continuing to divide and proliferate by replicating their genes without stopping the cell cycle, overcoming factors that promote cellular aging or limit replication, and avoiding the activation of the suicide genes that the cells possess themselves. Of course, recent research has enabled to change in the inherent properties of cells by manipulating genes, causing them to stop proliferating or die, or, conversely, to immortalize cells that have limited proliferation or division capabilities. Nevertheless, a significant portion of the cell lines currently in use are cancer cells/tumor cells, and these are largely due to the immortalization ability that the original cells possessed.

1.1.2 Tissue Cell Culture Techniques

Various cell culture techniques have been attempted since the early 1900s, but it was only when researchers began to fully handle cells in laboratories that they realized the importance of strict aseptic technique [5] and the subculture technique for adherent cells using trypsin [6]. Furthermore, in the 1940s and 1950s, it became possible to reduce bacterial contamination by adding antibiotics such as penicillin and streptomycin to the culture system [7, 8]. On the other hand, while the use of antibiotics in cell culture is useful for preventing microbial contamination, the point that they should be used at concentrations not affecting cell growth or other properties was already stressed in the 1950s, when cell culture became popular [8]. The addition of antibiotics and other drugs to cell culture systems can have undesirable effects, as evidenced by their significant changes in gene expression [9], and their impact must be considered with a full understanding of what is being assayed in the culture system.

1.1.3 Establishment of Cell Lines

Early results of cell culture include the success of continuous culture of mouse fibroblast L cells achieved in 1943 [10], and Sanford established the L929 cell line from these L cells in 1948 [11]. Starting with the establishment of the L929 cell line, HeLa cells (human cancer cell line) were established in 1952, and many other cell lines were subsequently established around 1960. This shows the enthusiasm for establishing cell lines at that time and also indicates that it was the time when various conditions for cell culture were being established. In fact, in the 1960s, a new method for dispersing tissue cells was proposed to compensate for the shortcomings of trypsin treatment [12].

Table 1.1 shows the cell lines that have been frequently used in the history of cell culture research, and it seems that almost all established cell lines used in general cell culture research, except for HeLa cells, are fibroblast-like cells. This is because fibroblasts produce collagen and other extracellular matrix proteins that serve as a scaffold for cells and can positively control the growth environment in a petri dish by promoting cell attachment and adhesion and are easy to separate from tissues by

Table 1.1 Cell lines that have been frequently used in cell culture research

Name of the cell line	Tissue of cell origin	Year established	Key references
L929	Mouse fibroblast	1948 (L cell had been originally established in 1943, but it was in 1948 that L929 was officially registered as a cell line)	Earle et al. [10]
HeLa	Human cervical cancer (HeLa is named after the patient, Henrietta Lacks)	1952	Gey et al. [2]
CHO	Chinese hamster ovary	1957	Puck et al. [13]
MDCK	Canine kidney from a normal, adult, female cocker spaniel (MDCK is an acronym of Madin–Darby canine kidney)	1958	1. ATCC [14] 2. Gaush et al. [15]
BHK-21	Syrian hamster kidney fibroblast	1961	Stoker and Macpherson [16]
Vero	African green monkey kidney	1962	Yasumura and Kawakita [17]
NIH3T3	Mouse embryonic fibroblast	1962	Todaro and Green [18]

Modified and created by the author based on the book of Magdalena Jedrzejczak-Silicka: History of Cell Culture (https://www.intechopen.com/chapters/53566). [FROM THE EDITED VOLUME, New Insights into Cell Culture Technology. Edited by Sivakumar Joghi Thatha Gowder]

the treatment with trypsin or other cell dispersing enzymes. Because fibroblasts are the cells that work to repair tissues and can switch from a quiescent state to a proliferation cycle depending on the tissue microenvironment, it can be said that they met the conditions for establishing cell lines, even with the level of cell culture technology at that time.

1.1.4 Research Using Cancer Cells

Another important purpose of cell culture is the establishment of cancer cell lines, the analysis of the cell biological functions of the obtained cells as tumors, the understanding of the mechanism of cancer growth and progression through the research about intracellular signals, and the use of knowledge from these analyses in the development of anticancer drugs to effectively treat cancer. To create a therapeutic strategy for cancer, a reliable preclinical model is necessary [19]. Thus, the establishment of good cell lines and the development of research methods using them are very important issues. Appropriate cell management, selection of proper research methods suitable to the purpose, and timely use of open-source cancer genome information, which has been rapidly developed in recent years, as well as facilities provided by various consortia, have paved the way for the better environment that can lead to accurate diagnostic and therapeutic methods for patients [20]. Taking these points into consideration, here the history of the use of cancer cell culture and its application to cancer research is briefly reviewed topic by topic.

1.1.4.1 Multidrug Resistance (MDR) of Cancer Cells

Attempts to determine the anticancer drug sensitivity of cancer cells using culture systems and apply it to clinical treatment have been made since the early days of cell culture technology. Among them, the establishment of anticancer drug-resistant cell strains provides a tool for the analysis of drug resistance mechanisms, the selection of effective therapeutic drugs, and the screening and development of new anticancer drugs. For example, the analysis of a cisplatin-resistant cell line (KFr) established from KF cells derived from human ovarian serous adenocarcinoma [21] has demonstrated that this cell line is effective in distinguishing anticancer drugs exhibiting cross-resistance from those not exhibiting and that the mechanism of cisplatin resistance is involved in the reduction of cisplatin accumulation in tumor cells [22]. There are many analytical studies on the mechanisms of resistance to individual anticancer drugs, so the explanation will not go into detail here, but particularly troublesome issues of multidrug resistance are mentioned.

The MDR gene, which confers multiple drug resistance to anticancer drugs, was first discovered in a study using Chinese hamster ovary (CHO) cells [23]. Subsequently, the human MDR gene was cloned from human KB cells, which exhibited multidrug resistance [24]. This gene encodes p-glycoprotein expressed on the cell membrane, and similar gene expression is also seen in human normal kidney cells and HepG2 liver cells. Subsequent studies have shown that resistance to anticancer drugs (multidrug resistance, MDR) is due to reduced drug accumulation caused by the expression of multiple ATP-dependent efflux pumps, and it has been pointed out that MDR1 gene polymorphisms in vivo may affect the pharmacokinetics of many commonly used drugs, including anticancer drugs [25]. Several factors, including genetics, microRNAs (miRNAs), and long noncoding RNAs (lncRNAs), are known to contribute to the development of such multidrug resistance in cancer cells. Therefore, it has been shown that certain miRNAs have value as biomarkers and may serve as a valuable source of information for the selection of optimal chemotherapy regimens in breast cancer patients [26]. Since the mechanism of such multidrug resistance is due to the inherent overexpression of ATP-binding cassette (ABC) transporter proteins on the cell membrane, strategies being considered include the use of effective inhibitors against these proteins, approaches such as knocking out or mRNA silencing of the MDR1 gene, alternative splicing of the responsible gene, and the use of nanotechnology to design drugs that are not recognized by the p-glycoprotein pump function [27].

1.1.4.2 Oncogenes and Tumor Suppressor Genes

Research about cell carcinogenesis, which began in the 1970s, led to the discovery of oncogenes and tumor suppressor genes. In the mid-1970s, Bishop and Varmus et al. discovered the world's first oncogene, *src* (c-Src), by analyzing the oncogenic mechanism of Rous sarcoma virus, a virus that is tumorigenic in birds. They also demonstrated that the oncogene *src* possessed by the Rous sarcoma virus is present in normal cells in the form of a protooncogene, proving for the first time the involvement of genes in the mechanism of carcinogenesis [28, 29]. At that time, genetic engineering was advancing, and the enzyme having an RNA-dependent DNA polymerase activity, named reverse transcriptase, that had been discovered in 1970 [30, 31] was used in this study.

In addition, to clone oncogenes from human bladder cancer cell line T24, NIH3T3 mouse fibroblasts were transfected with DNA derived from T24 cells, and the *ras* gene was identified from the human portion of DNA extracted from transformed NIH3T3 cell masses that had lost contact inhibition and showed neoplastic growth [32, 33]. These breakthrough studies led to the knowledge of the existence and role of cancer genes, and with the maturation of molecular cloning technology, many cancer genes were subsequently cloned.

Meanwhile, the first tumor suppressor gene was isolated from the studies using retinoblastoma, in which a deletion in the long arm of chromosome 13 was known. In 1986, Dryja et al. identified a loss of function in one of the two homologous copies of *Rb*, a single locus that maps to band q14 of human chromosome 13 [34]. In the following year, the base sequence of the gene *Rb* was determined [35].

p53 is a well-known tumor suppressor gene. When it was discovered in 1979, it was classified as an oncogene because of its high expression in tumor cells. It was later discovered that this was due to the dominant negative expression of mutant (loss-of-function) forms of p53 molecules; therefore, it was reclassified as a tumor suppressor gene. The protein p53 encoded by the *TP53* gene is called the "guardian of the genome" [36], which is closely related to cell cycle control and apoptosis control. Since then, new tumor suppressor genes have been discovered one after another, and their relationship to cell proliferation and cell cycle control has been studied.

The discovery of these oncogenes and tumor suppressor genes led to the analysis of how intracellular signals are involved in the proliferation and progression of cancer cells isolated from clinical specimens and what kind of anticancer drug they could be targeted by. This led to the creation of many cell lines derived from tumor cells, which have established their position as tools for the development of new anticancer drugs and molecular targeted drugs.

1.1.4.3 Cancer Genomics/Translational Research

The progress of translational research since the 2000s has been largely due to the ability to directly evaluate clinical samples. For this reason, establishing preclinical models to verify therapeutic strategies is a key issue to promote actual applications [19]. Numerous cancer cell lines have been established, and knowledge has been accumulated through numerous culture experiments. Many in vitro and in vivo models have also been proposed. However, it has been pointed out that these may not necessarily be well connected to actual clinical diagnosis and subsequent treatment proposals. While the verification of the mechanism of anticancer drug activity against cancer cells and integration of such pharmacological knowledge have not been achieved enough, one that is particularly eye-catching is the National Cancer Institute's 60 Human Tumor Cell Line Anticancer Drug Screening (NCI60) (https://dtp.cancer.gov/discovery_development/nci-60/), which was developed in the late 1980s as an in vitro drug discovery tool aimed at replacing animal tumor transplant experiments in anticancer drug screening. This now plays a role as a service screen to support the cancer research community [37]. For over 30 years, NCI60 has been screening and assaying numerous compounds as a drug discovery and development researcher for anticancer drugs, but since October 2023, it has been changing to a new platform (automated high-throughput assay system).

Despite these efforts, the integration of pharmacological knowledge of anticancer drugs against cancer cells is still slow. Panel analysis of cancer cell lines was practically limited to a maximum of 60 cell lines at NCI60. When the Cancer Genome Anatomy Project (CGAP) began efforts to define the genetic basis of human cancer (currently it exists in a different form: https://mitelmandatabase.isb-cgc.org/), it was realized that a similar effort was needed to characterize cancer cell lines, and so the Cancer Cell Line Encyclopedia (CCLE) was created (https://sites.broadinstitute.org/ccle/). This CCLE has accelerated the movement toward predictive modeling of anticancer drug sensitivity of cancer cells [38].

Until around 2000, cancer research, backboned by molecular biology, promoted genetics as an analytical method to understand the tumor properties, i.e., proliferation, invasiveness, cell death avoidance, angiogenesis, metastatic potential, etc., at the genetic level by analyzing the genes of cancer cells. This was aimed at finding characteristic genetic changes in individual cancer cells and analyzing them in detail to understand their growth and progression patterns, thereby searching for molecular targets in which anticancer drug sensitivity is well drawn as a point of therapeutic intervention. This effort not only led to the discovery of many cancer genes and tumor suppressor genes but also to the development of newly designed drugs that target them. Typical examples of the earliest successful molecular targeted drugs are trastuzumab (Herceptin: monoclonal antibody against HER2 protein) [39] and imatinib (Gleevec: Abelson tyrosine kinase inhibitor) [40]. This has significantly improved the treatment outcomes of breast cancer and chronic myeloid leukemia, and many molecular-targeted drugs have been developed since then.

However, the limitations of doing research only in the field of genetics are becoming apparent. In other words, cancer is a multifactorial disease, and therefore developing treatments that target only specific genes seems not necessarily successful. The omics approach has been introduced as a new direction to overcome these barriers. Omics is an approach to collectively analyze the life activities of cells, and genomics, which collectively analyzes genes, and proteomics, which collectively analyzes protein expression, have come to be used. In particular, the progress in cancer genomics has been remarkable, and it has served as a foundation for a better understanding of the molecular basis of cancer proliferation, metastasis, and drug resistance [41]. Over the past decade, omics research has expanded beyond genomics and proteomics to metabolomics, transcriptomics, metagenomics, epigenomics, and more. Combined with the development of high-throughput sequencing technology for DNA and RNA, as well as information and communication technology (ICT) and artificial intelligence (AI) technology, this has led to a detailed understanding of cellular activity through the analysis of large amounts of data and to its use in large-scale research and epidemiological clinical trials. As an example, efforts have begun to accumulate and analyze data on a series of cancer genomes to utilize them in multidisciplinary cancer treatment. In the United States, the Cancer Genome Atlas (TCGA) program (https://www.cancer.gov/ccg/research/genome-sequencing/tcga), a joint research project between the National Cancer Institute (NCI) and the National Human Genome Research Institute, began in 2006 and has generated genome, epigenome, transcriptome, and proteome data for many cancer cell lines, which are already helping to improve the ability to diagnose, treat, and prevent cancer.

1.1.5 Stem Cell Research

Stem cell research using cell culture techniques is roughly divided into two categories. One is to understand the mechanisms of development from stem cells to the individual level and differentiation into functional cells and living organs, leading to a broader understanding of life activities as academic knowledge, namely, the foundation of biology. Second is research activities which include a practical aspect of utilizing such knowledge in assisted reproductive medicine and regenerative medicine.

Early stem cell research attempted to understand hematopoietic stem cell differentiation and lineage formation through colony-forming assays, where many growth factors and colony-stimulating factors were isolated and identified through such studies. In addition, the involvement of gene expression and growth factors required for differentiation in immune and lymphatic tissues, the mechanism of lymphocyte maturation and DNA rearrangement by recombinant gene expression, and the mechanism of class switching in antibody gene expression have been elucidated one after another. Such studies can also be applied to investigating the efficiency of bone marrow transplantation after radiation exposure and to confirm the differentiation of hematopoietic cells from transplanted cells. In recent years, a new method [42] has been developed to confirm the clonality of cells using a small fraction of single-cell suspensions by improving the traditional clonogenic assay [43].

On the other hand, mesenchymal stem cells (MSCs) have been attracting attention in recent years from the perspective of regenerative medicine. MSCs are mesenchymal stromal cells that have the pluripotency to differentiate into various cells such as bone, cartilage, muscle, and fat. Due to their relative ease of collection and the availability of sufficient numbers of cells, many attempts have been made to use MSCs as a starting point to induce differentiation into desired cell populations using various stimuli [44]. MSCs are becoming a central material in tissue engineering, and many clinical trials have already been conducted. However, treatments aimed at tissue replacement and repair using MSCs are not established clinical procedures, so ethical, economic, and legal aspects must be considered when implementing them.

Stem cell research also occupies an important position in the field of cancer research in cell culture. Since some cancer cells can self-replicate, differentiate, and proliferate, they are called cancer stem cells. The concept of cancer stem cells has existed for a long time, but it was Bonnet and Dick's study on acute myeloid leukemia cells that demonstrated it in an easy-to-understand way [45], which showed the existence of a source cell that developed into leukemia cells. In cancer radiation therapy and chemotherapy, the presence of cancer stem cells within the remaining treatment-resistant cells is believed to make a complete cure difficult, thereby being the main source of recurrence and metastasis. The presence of cancer stem cells brings about heterogeneity in cancer tissue, and the frequent presence of cells from different tissue systems in clinical cancers may support this evidence.

Research on the early development of humans has been subjected to ethical restrictions, so experiments involving the manipulation and analysis of early embryos have traditionally used animals or animal cells. In 1981, mouse embryonic stem (ES) cells were established [46, 47], which brought a great advance in tracking the development of embryos from fertilized eggs. Furthermore, the development of recombinant DNA technology and gene manipulation technology strongly assisted the creation of transgenic and knockout mice, which allowed us to observe the phenotype of an embryo when the genes of fertilized eggs were manipulated. This was a major advance in the field of genetics, and our understanding of genes and their phenotypes rapidly deepened.

Regarding the research using human-fertilized eggs, regulations vary by country and region. Some countries allow the use of surplus embryos, i.e., fertilized embryos created for fertility treatment that will be discarded without being used for implantation, in the experiments under the knowledge and consent of the donor. Therefore, human embryonic stem cell lines have been established from these embryos, making it possible to trace cell biological events in early human development, at least up to the stage before implantation. Genome editing technology using the CRISPR/Cas9 system, discovered in 2012 [48], is easy, inexpensive, and efficient and is now being used to manipulate the genome of human embryos. In 2018, the WHO established a global, multidisciplinary Expert Advisory Committee to consider scientific, ethical, social, and legal issues related to human genome editing and in 2021 published a governance framework and recommendations for human genome editing (https://www.who.int/health-topics/human-genome-editing/#tab=tab_3). In addition, the Human Genome Editing (HGE) Registry was also established (https://www.who.int/groups/expert-advisory-committee-on-developing-global-standards-for-governance-and-oversight-of-human-genome-editing/registry).

Based on the extraordinary idea of artificially creating the pluripotency seen in embryonic stem cells, Professor Shinya Yamanaka created induced pluripotent stem (iPS) cells in 2006 [49]. By using the reverse idea of transforming normal cells with four genes (Oct3/4, Sox2, Klf4, and c-Myc) that are significantly expressed in embryonic stem cells, they demonstrated that pluripotent stem cells can be induced from normal differentiated cells. This was an experiment in murine cells, but the following year they succeeded in producing human iPS cells [50]. The emergence of iPS cells has brought great progress in the research field of the early differentiation of human cells, many of which had been ethically difficult. In addition, due to their convenience, they are being used to examine drug sensitivity in disease models and apply them to regenerative medicine. It has already become possible to generate primordial germ cell-like cells from iPS cells [51]. Also, the effort to elucidate the framework of epigenetic reprogramming in human cells has been made to develop in vitro gametogenesis (IVG) [52]. It is expected that such progress in research will lead to the elucidation of the mystery of why only germline cells can pass on life to the next generation.

1.1.6 Industrial Use of Cell Lines

Cultivation of the cells is essential to grow the viruses, and many established cell lines have been used for this purpose. Although cell culture technology has made a great contribution to this field, some problematic events have also occurred. For example, some of the polio vaccines administered between 1955 and 1963 were contaminated with simian virus 40 (SV40) [53]. SV40 is a carcinogenic virus, and although there have been no reports of cancer caused by those contaminated vaccines, eliminating contamination has become an important issue [54]. Currently, cell lines such as Vero cells, Madin-Darby canine kidney (MDCK) cells, or chicken embryo fibroblasts (CEFs) are often used in vaccine development, but attempts have been made to establish the so-called designer cells that are safe and meet the essential requirements [55].

Various cell lines have been used for the development of biopharmaceuticals such as recombinant proteins and antibody drugs. In recent years, there has been a trend toward using human cells instead of animal cells such as CHO cells and mouse myeloma cells, which have been used in the past. The main reason for this is that the resulting recombinant proteins are more likely to have posttranslational modifications that match those found in endogenous human proteins [56]. In the future, it will be desirable to thoroughly check for viral contamination specific to human cells and to use cells optimized for the intended purpose.

The creation of recombinant viruses using human cells is being actively applied to medical purposes such as gene therapy. For example, the delivery system using adeno-associated viruses, which are known to be specifically integrated into human chromosome 19, is established as a new gene therapy tool with its safety and high gene expression efficiency [57]. HEK239 cells or HEK239T cells in suspension culture are used to produce those viruses. The HEK239 cells, frequently used for genetic engineering, are human embryonic kidney cells transformed with DNA fragments from adenovirus type 5 [58], and this cell line is frequently used as a platform for producing recombinant proteins and viral vectors. To date, there have been many attempts to develop biopharmaceuticals using animal-derived cell lines such as CHO cells. In contrast, one of the key advantages of using human cell lines such as HEK239 cells is that the resulting biopharmaceuticals are likely to have more "human-like" posttranslational modifications, such as proper gamma-carboxylation of glutamic acid and sulfation of tyrosine residues, as well as better reproducibility of human glycan structures during recombinant protein production [59]. However, thanks to improvements in genetic engineering and genome editing technologies, it is now possible to design CHO cells in a form that is favorable for human drug development. Taking advantage of the accumulated know-how, an environment suitable for the development of biopharmaceuticals is being created.

1.2 Basics of Cell Culture and Related Technologies

From Chap. 2 onwards, for readers who will be reading about new findings and new research attempts on cell culture, here we would like to first explain some points to note and basic points to understand about cell culture. To that end, the terms frequently used in cell culture are summarized in Table 1.2. Also, important cells and technologies in the history of cell culture are listed in Table 1.3.

1.3 Cell Authentication

Many public cell banks have already been established to support research, and the research community is making effective use of them to advance research. Researchers also share research samples and data to expand the scope of their research activities. However, there are major pitfalls in using these cells that researchers are unaware of. We researchers cannot ignore the issue of cell authentication to properly conduct experimental research using cultured cells [67].

Microbial contamination of cultured cells not only affects cell viability and proliferation but also affects gene expression and changes the phenotype of the cells. This situation not only makes it difficult to compare the results of previous experiments using the cells but also undermines the reliability of the experimental results themselves and reduces the applicability of the cells to various analyses. In addition

Table 1.2 Terms frequently used for cell culture

Term	Definition	Importance/application
Anchorage dependence/ independence	Anchorage-dependency is the cell nature of adhesiveness required for proliferation. Anchorage-independent cells do not require adhesiveness for cell proliferation, and cells can proliferate in suspension	To increase cell adhesion, extracellular matrix (ECM) is solidified on the culture surface of the culture dish
Authentication	Proof that cell lines are of the same origin	Authentication ensures that the results of experiments can be compared with and referenced to other experiments using the same cell line. The authenticity of the cells is required when publishing research papers
Cell line	The descendants of cells that are produced when established continuously growing cells are subcultured	If a cell line can be created, it can be applied to a wide range of research topics. However, when using cells that cannot be created into a cell line, it is difficult to prove the reproducibility of the research
Clone	This refers to cells that can be isolated from colonies derived from a single cell. The process of isolating such cells is called cell cloning	Because the cloned cells have the same genetic background, they are thought to have the same phenotype and characteristics
Confluence	A condition in which all cells are in contact with each other, leaving no room for cell proliferation	The time to reach confluence is an indicator of the proliferation ability of the cells and the estimated time required for the next subculture
Contact inhibition	Cessation of cellular movement, growth, and division upon contact with other cells	The presence or absence of contact inhibition is an indicator for determining whether the cells have the property of tumor proliferation
Contamination	A state in which cultured cells become contaminated with microorganisms, impairing their optimal growth. Or, cultured cells become mixed with other cells (=cross contamination)	Microbial contamination significantly inhibits cell growth and threatens cell survival. It also significantly spoils experimental results. Mycoplasma contamination is common in clinical samples and often goes unnoticed for a short period of cultivation time. In studies using cells suspected of cross contamination, doubts arise about the interpretation of experimental results
Differentiation	Differentiation is the process by which an immature cell acquires the properties of a cell with a specific function	In general, once differentiated into specific functional cells, normal cells do not revert to their original state. Reversion to their original function/appearance is called dedifferentiation, and such properties can be observed in tumor cells

(continued)

Table 1.2 (continued)

Term	Definition	Importance/application
Doubling time	Time required for the cells to double in number, assuming that the cells continued to grow at a constant rate	Doubling time is an indicator of cell proliferation speed and, in the case of tumors, of their malignancy
Generation number	Number of times the cell line population has doubled since the cells were isolated	This is an index of how many passages an established cell line has undergone. Since some cells change in properties with each passage, this value is an important factor for standardizing experimental conditions
Genotype	Genotype is the genetic constitution of an individual cell	Advances in genetic engineering technology have enabled the manipulation of genotypes
Immortalization	The acquisition of the ability for cells to grow and be subcultured indefinitely	Cells can be immortalized by introducing virus-derived genes and a telomerase gene. Some tumor cells have the innate ability to become immortal
Passage	Passing cultured cells to the next culture	It is important to maintain the properties of the original cells during passage. To achieve this, cell culture conditions must be strictly defined, and the prevention of contamination is also an important point
Phenotype	Phenotype is observable properties of the cells that result from the interaction of genotype and culture environment	It is possible to identify the causative gene from the difference in phenotype
Primary culture	Refers to the first cell culture isolated from a sample	This is often the first step in establishing cell lines from clinical samples
Subculture	Transferring cultured cells to a different culture system	This is a critical step in cell cloning, not just in transferring the cells to another cell culture system
Senescence	A process resulting from various stresses, in which cells undergo irreversible growth arrest (known as cellular senescence, as opposed to individual aging)	Cellular senescence is associated with the expression of characteristic proteins (p53/p21 and $p16^{Ink4a}$) related to DNA damage and brings the proliferation limit (Hayflick limit)
Spheroid	Self-organized spheroidal cell aggregates, which are induced by creating an environment that prevents cells from adhesion to the flat culture dish bottom	Although two-dimensional monolayer culture is simple and often used in experiments, it is not a natural cell growth environment. Therefore, three-dimensional spheroid culture is used as a culture method closer to the in vivo environment

(continued)

Table 1.2 (continued)

Term	Definition	Importance/application
Transformation	A hereditary change that involves a change in genotype. It can occur naturally, but can also be manipulated artificially. For your reference, the term "transfection" is a combination of the words transformation and infection	This is the word often used to explain the altered growth characteristics associated with malignant changes, such as anchorage independence, loss of contact inhibition, evasion from density restriction of cell proliferation, and tumor formation in vivo
Tumorigenesis	The ability of tissues or cells to form tumors when transplanted in vivo	This is an indicator of whether the cells have tumorigenic potential

Modified citation with reference to "Geraghty RJ, Capes-Davis A, Davis JM, Downward J, Freshney RI, Knezevic I, et al. Cancer Research UK. Guidelines for the use of cell lines in biomedical research. *Br J Cancer.* 2014; **111**:1021–46"

to microbial contamination by mycoplasma, cross-contamination, such as mistaking a cell line for another or mixing it with another cell line, can greatly change the results of experimental research, leading to the spread of erroneous research results without recognizing it, and unwanted influence on the research results of other laboratories. HeLa cells are the oldest established cell line and have been used for research for the longest time and in most laboratories. However, they are also the cells that have been contaminated the most frequently. For example, it became clear that many cancer cell lines used to model various types of cancer were derived from the HeLa cervical carcinoma line, and not from other cancers [68], and chromosomal analysis of many cell samples showed contamination patterns characteristic of HeLa [69].

The problem of insufficient cell authentication, together with problems with microbial contamination of cells and cell handling techniques such as isolation, culture, and subculture, greatly reduces the value of experimental research. To address these issues, the United Kingdom Coordinating Committee on Cancer Research (UKCCCR) established guidelines in 2000 that stipulate the appropriate handling of cells [70]. Furthermore, cell usage guidelines have been revised and published to be accessible to researchers not only in the United Kingdom but also around the world [71]. To specifically address the issue of cell line misidentification, standards for the authentication of human cell lines using short tandem repeat profiling have also been developed [72].

It is very difficult for researchers to check for cross-contaminated or misidentified cell lines by themselves. To answer this question, a list of known cross-contaminated cell lines was compiled by a group of volunteer researchers in 2010 [73]. At the time, the list included 360 cell lines extracted from 68 references. Most of these contaminations occurred within the same species, and HeLa remains the most frequently contaminated human cell line. To address this issue, the International Cell Line Authentication Committee (ICLAC) (https://iclac.org/) was organized in 2012 and has been providing a database on authentication. It has also released the

Table 1.3 Important cells and technologies in the history of cell culture

Cells and technologies	Year of established or invented	Significance	Key references
L929 cells	1943	The first cell line established in the history of cell culture	Earle et al. [10]
HeLa cells	1951	The first established human cancer cell line	Gey et al. [2]
Use of dimethyl sulphoxide (DMSO) in cell cryopreservation	1959	Reduce cell damage caused when freezing and storing cells	Lovelock and Bishop [60]
Microinjection	1966	Although the basic technique had been around since the early nineteenth century, this was the first time a known liquid was injected into a fertilized mouse egg (a technique used to create transgenic mice)	Lin [61]
Calcium phosphate method	1973	Shown to be an efficient method for transformation	Graham and van der Eb [62]
Hybridoma	1975	A method of fusing myeloma cells with antibody-producing cells to produce large amounts of monoclonal antibodies specific to a particular antigen	Köhler and Milstein [63]
HEK239 cells (human embryonic kidney cells)	1977	A platform for producing recombinant proteins and viral vectors	Graham et al. [58]
Embryonic stem cells	1981	The establishment of pluripotent stem cells from mouse embryos that can develop into whole bodies has greatly advanced research about early development	1. Evans and Kaufman [46] 2. Martin [47]
Lipofection method	1987	5 to 100 times more effective than calcium phosphate or DEAE-dextran transfection methods	Felgner et al. [64]
RNA interference (RNAi)	1998	Gene silencing by double-stranded RNA (short interfering RNA: siRNA)	Fire et al. [65].
iPS cells	2006	Pluripotent stem cells can be induced from normal differentiated cells by introducing four genes (Oct3/4, Sox2, Klf4, and c-Myc). iPS cells can also be produced in humans (2007)	1. Takahashi and Yamanaka [49] 2. Takahashi et al. [50]
Genome editing technology (CRISPR/Cas9)	2012	A technology that can cut and edit genomic DNA in a sequence-specific manner. This is possible by simply introducing guide RNA that recognizes the sequence and an enzyme (Cas9) that cuts DNA into living cells	1. Jinek et al. [48] 2. Doudna and Charpentier [66]

Guide to Human Cell Line Authentication (https://iclac.org/wp-content/uploads/ICLAC_Guide-to-Human-Cell-Line-Authentication_02-Mar-2023.pdf).

1.4 Ethical Issues in Sequencing the HeLa Genome

In 2013, the haplotype-resolved whole-genome sequencing of the HeLa CCL-2 strain was completed, providing a high-quality reference genome for HeLa cells for future research activities [74]. These tumor cells were isolated and immortalized from a cancer patient, Henrietta Lacks, more than 60 years ago, and the ethical standards at the time did not require the patient and her family to consent to the use of the cells in research. However, the issue was left as it was, and many laboratories around the world frequently used these cells for research without this "informed consent."

If it had been cell culture experiments or drug development research using them, this issue might have remained on the back burner, but when the research progressed to genomic analysis of tumor cells, the issue of personal information came to light. In other words, through the analysis of the genomic information of these tumor cells, the genome of Henrietta Lacks and her family, that is, the ultimate personal information, would be made public. Naturally, the process of Henrietta Lacks' family's consent to the sequencing of the HeLa cell genome and its use in research had never been taken, so ethical issues arose. Eventually, a meeting was held between members of the Lacks family and the leadership of the National Institutes of Health (NIH), and an agreement was reached under which all publications based on HeLa cell research would include a letter of thanks to the Lacks family, and the Lacks family would comply with requests for use of HeLa cell genomic data [75]. This issue clearly illustrates the ethical issues that must be considered before proceeding with research using cell lines. And hopefully, a harmonious approach has been taken to resolve them.

Concerning this issue, 2020 marks the 100th birthday of Henrietta Lacks, and Collins, who was the director of the NIH, and others have reconsidered the need for research participants to be independent [76]. HeLa cells are the cell line that is used in most laboratories around the world and have contributed to numerous medical advances. However, the fact that they were used without the permission of Henrietta and her family, given that she was a black woman, and overlapping with past racist research activities such as the Tuskegee syphilis experiment, has once again triggered consideration of ethical issues.

1.5 The Way Forward

Cell culture technology has more than 100 years of accumulated knowledge and achievements. In the early days, various methods were tested by trial and error, and the most suitable one was selected. The way of doing research, i.e., trial and error, may still be the way it is today. On the other hand, we have already gained a lot of knowledge about the mechanism of cell division and proliferation, gene expression, and cell metabolism related to it. We have also reached the point where we can freely manipulate the phenotype of cells by manipulating the genome. The goal of our research using cell culture is to understand the mysteries of life through cell culture technology, elucidate the mechanisms of disease, and apply it to industry and medicine. In the future, the experimental conditions may shift significantly from wet experiments to research using ICT and AI. However, the most important thing is to properly determine the true purpose of the research and perform it with highly transparent manners by considering various ethical issues.

References

1. Nott R. HeLa cell line. Embryo project encyclopedia. Tempe: Arizona State University; 2020. https://embryo.asu.edu/pages/hela-cell-line.
2. Gey GO, Coffman WD, Kubicek MT. Tissue culture studies of the proliferative capacity of cervical carcinoma and normal epithelium. Cancer Res. 1952;12:264–5.
3. Seedorf K, Oltersdorf T, Krammer G, Rowekamp W. Identification of early proteins of the human papilloma viruses type 16 (HPV 16) and type 18 (HPV 18) in cervical carcinoma cells. EMBO J. 1987;6:139–44.
4. Castro-Muñoz LJ, Manzo-Merino J, Muñoz-Bello JO, Olmedo-Nieva L, Cedro-Tanda A, Alfaro-Ruiz LA, et al. The human papillomavirus (HPV) E1 protein regulates the expression of cellular genes involved in immune response. Sci Rep. 2019;9:13620.
5. Carrel A, Burrow MT. Cultivation of tissues in vitro and its technique. J Exp Med. 1911;13:387–96.
6. Rous P, Jones FS. A method for obtaining suspensions of living cells from the fixed tissues, and for the plating out of individual cells. J Exp Med. 1916;23:549–55.
7. Keilov H. The effect of streptomycin on tissue cultures. Experientia. 1948;4:483–4.
8. Cruickshank CN, Lowbury EJ. Effect of antibiotics on tissue cultures of human skin. Br Med J. 1952;2:1070–2.
9. Ryu AH, Eckalbar WL, Kreimer A, Yosef N, Ahituv N. Use antibiotics in cell culture with caution: genome-wide identification of antibiotic-induced changes in gene expression and regulation. Sci Rep. 2017;7:7533.
10. Earle WR, Schilling EL, Stark TH, Straus NP, Brown MF, Shelton E. Production of malignancy in vitro. IV. The mouse fibroblast cultures and changes seen in the living cells. J Natl Cancer Inst. 1943;4:165–212.
11. Sanford KK, Earle WR, Likely GD. The growth in vitro of single isolated tissue cells. J Natl Cancer Inst. 1948;9:229–46.
12. Gwatkin R, Thomson J. A new method for dispersing the cells of mammalian tissues. Nature. 1964;201:1242–3.
13. Puck TT, Cieciura SJ, Robinson A. Genetics of somatic mammalian cells. III. Long-term cultivation of euploid cells from human and animal subjects. J Exp Med. 1958;108:945–56.

14. ATCC. https://www.atcc.org/products/ccl-34#characteristics.
15. Gaush CR, Hard WL, Smith TF. Characterization of an established line of canine kidney cells (MDCK). Exp Biol Med. 1966;122:931–5.
16. Stoker M, Macpherson I. Syrian hamster fibroblast cell line *BHK*21 and its derivatives. Nature. 1964;203:1355–7.
17. Yasumura Y, Kawakita Y. Vero cell line derived from the kidney of a normal, adult, African green monkey (Cercopithecus). Nippon Rinsho. 1963;21:1209.
18. Todaro GJ, Green H. Quantitative studies of the growth of mouse embryo cells in culture and their development into established lines. J Cell Biol. 1963;17:299–313.
19. Gillet JP, Varma S, Gottesman MM. The clinical relevance of cancer cell lines. J Natl Cancer Inst. 2013;105:452–8.
20. Mirabelli P, Coppola L, Salvatore M. Cancer cell lines are useful model systems for medical research. Cancers (Basel). 2019;11:1098.
21. Kikuchi Y, Miyauchi M, Kizawa I, Oomori K, Kato K. Establishment of a cisplatin-resistant human ovarian cancer cell line. J Natl Cancer Inst. 1986;77:1181–5.
22. Kikuchi Y, Iwano I, Miyauchi M, Kita T, Oomori K, Kizawa I, et al. The mechanism of acquired resistance to cisplatin by a human ovarian cancer cell line. Jpn J Cancer Res. 1988;79:632–5.
23. Juliano RL, Ling V. A surface glycoprotein modulating drug permeability in Chinese hamster ovary cell mutants. Biochim Biophys Acta. 1976;455:152–62.
24. Ueda K, Clark DP, Chen CJ, Roninson IB, Gottesman MM, Pastan I. The human multidrug resistance (mdr1) gene. cDNA cloning and transcription initiation. J Biol Chem. 1987;262:505–8.
25. Ambudkar SV, Kimchi-Sarfaty C, Sauna ZE, Gottesman MM. P-glycoprotein: from genomics to mechanism. Oncogene. 2003;22:7468–85.
26. Chen X, Lu P, Wang DD, Yang SJ, Wu Y, Shen HY, et al. The role of miRNAs in drug resistance and prognosis of breast cancer formalin-fixed paraffin-embedded tissues. Gene. 2016;595:221–6.
27. Duan C, Yu M, Xu J, Li BY, Zhao Y, Kankala RK. Overcoming cancer multi-drug resistance (MDR): reasons, mechanisms, nanotherapeutic solutions, and challenges. Biomed Pharmacother. 2023;162:114643.
28. Stehelin D, Varmus HE, Bishop JM, Vogt PK. DNA related to the transforming gene(s) of avian sarcoma viruses is present in normal avian DNA. Nature. 1976;260:170–3.
29. Oppermann H, Levinson AD, Varmus HE, Levintow L, Bishop JM. Uninfected vertebrate cells contain a protein that is closely related to the product of the avian sarcoma virus transforming gene (src). Proc Natl Acad Sci U S A. 1979;76:1804–8.
30. Baltimore D. RNA-dependent DNA polymerase in virions of RNA tumour viruses. Nature. 1970;226:1209–11.
31. Temin HM, Mizutani S. RNA-dependent DNA polymerase in virions of Rous sarcoma virus. Nature. 1970;226:1211–3.
32. Shih C, Weinberg RA. Isolation of a transforming sequence from a human bladder carcinoma cell line. Cell. 1982;29:161–9.
33. Parada LF, Tabin CJ, Shih C, Weinberg RA. Human EJ bladder carcinoma oncogene is homologue of Harvey sarcoma virus *ras* gene. Nature. 1982;297:474–8.
34. Friend SH, Bernards R, Rogelj S, Weinberg RA, Rapaport JM, Albert DM, et al. A human DNA segment with properties of the gene that predisposes to retinoblastoma and osteosarcoma. Nature. 1986;323:643–6.
35. Lee WH, Bookstein R, Hong F, Young LJ, Shew JY, Lee EY. Human retinoblastoma susceptibility gene: cloning, identification, and sequence. Science. 1987;235:1394–9.
36. Lane DP. Cancer. p53, guardian of the genome. Nature. 1992;358:15–6.
37. Shoemaker RH. The NCI60 human tumour cell line anticancer drug screen. Nat Rev Cancer. 2006;6:813–23.

38. Barretina J, Caponigro G, Stransky N, Venkatesan K, Margolin AA, Kim S, et al. The cancer cell line encyclopedia enables predictive modelling of anticancer drug sensitivity. Nature. 2012;483:603–7.
39. Slamon DJ, Clark GM, Wong SG, Levin WJ, Ullrich A, McGuire WL. Human breast cancer: correlation of relapse and survival with amplification of the HER-2/neu oncogene. Science. 1987;235:177–82.
40. Braun TP, Eide CA, Druker BJ. Response and resistance to BCR-ABL1-targeted therapies. Cancer Cell. 2020;37:530–42.
41. Center for Cancer Genomics, National Cancer Institute. Cancer genomics overview. https://www.cancer.gov/ccg/research/cancer-genomics-overview.
42. Brix N, Samaga D, Belka C, Zitzelsberger H, Lauber K. Analysis of clonogenic growth in vitro. Nat Protoc. 2021;16:4963–91.
43. Puck TT, Morkovin D, Marcus PI, Cieciura SJ. Action of X-rays on mammalian cells. II. Survival curves of cells from normal human tissues. J Exp Med. 1957;106:485–500.
44. Pittenger MF, Discher DE, Péault BM, Phinney DG, Hare JM, Caplan AI. Mesenchymal stem cell perspective: cell biology to clinical progress. NPJ Regen Med. 2019;4:22.
45. Bonnet D, Dick JE. Human acute myeloid leukemia is organized as a hierarchy that originates from a primitive hematopoietic cell. Nat Med. 1997;3:730–7.
46. Evans MJ, Kaufman MH. Establishment in culture of pluripotential cells from mouse embryos. Nature. 1981;292:154–6.
47. Martin GR. Isolation of a pluripotent cell line from early mouse embryos cultured in medium conditioned by teratocarcinoma stem cells. Proc Natl Acad Sci U S A. 1981;78:7634–8.
48. Jinek M, Chylinski K, Fonfara I, Hauer M, Doudna JA, Charpentier E. A programmable dual-RNA-guided DNA endonuclease in adaptive bacterial immunity. Science. 2012;337:816–21.
49. Takahashi K, Yamanaka S. Induction of pluripotent stem cells from mouse embryonic and adult fibroblast cultures by defined factors. Cell. 2006;126:663–76.
50. Takahashi K, Tanabe K, Ohnuki M, Narita M, Ichisaka T, Tomoda K, et al. Induction of pluripotent stem cells from adult human fibroblasts by defined factors. Cell. 2007;131:861–72.
51. Mitsunaga S, Shioda K, Hanna JH, Isselbacher KJ, Shioda T. Production and analysis of human primordial germ cell-like cells. Methods Mol Biol. 2021;2195:125–45.
52. Murase Y, Yokogawa R, Yabuta Y, Nagano M, Katou Y, Mizuyama M, et al. In vitro reconstitution of epigenetic reprogramming in the human germ line. Nature. 2024;631:170.
53. Institute of Medicine (US) Immunization Safety Review Committee. In: Stratton K, Almario DA, McCormick MC, editors. Immunization safety review: SV40 contamination of polio vaccine and cancer. Washington, DC: National Academies Press (US); 2002.
54. Dang-Tan T, Mahmud SM, Puntoni R, Franco EL. Polio vaccines, simian virus 40, and human cancer: the epidemiologic evidence for a causal association. Oncogene. 2004;23:6535–40.
55. Genzel Y. Designing cell lines for viral vaccine production: where do we stand? Biotechnol J. 2015;10:728–40.
56. Dumont J, Euwart D, Mei B, Estes S, Kshirsagar R. Human cell lines for biopharmaceutical manufacturing: history, status, and future perspectives. Crit Rev Biotechnol. 2016;36:1110–22.
57. Kimura T, Ferran B, Tsukahara Y, Shang Q, Desai S, Fedoce A, et al. Production of adeno-associated virus vectors for in vitro and in vivo applications. Sci Rep. 2019;9:13601.
58. Graham FL, Smiley J, Russell WC, Nairn R. Characteristics of a human cell line transformed by DNA from human adenovirus type 5. J Gen Virol. 1977;36:59–74.
59. Tan E, Chin CSH, Lim ZFS, Ng SK. HEK293 cell line as a platform to produce recombinant proteins and viral vectors. Front Bioeng Biotechnol. 2021;9:796991.
60. Lovelock J, Bishop M. Prevention of freezing damage to living cells by dimethyl sulphoxide. Nature. 1959;183:1394–5.
61. Lin TP. Microinjection of mouse eggs. Science. 1966;151:333–7.
62. Graham FL, van der Eb AJ. A new technique for the assay of infectivity of human adenovirus 5 DNA. Virology. 1973;52:456–67.

63. Köhler G, Milstein C. Continuous cultures of fused cells secreting antibody of predefined specificity. Nature. 1975;256:495–7.
64. Felgner PL, Gadek TR, Holm M, Roman R, Chan HW, Wenz M, et al. Lipofection: a highly efficient, lipid-mediated DNA-transfection procedure. Proc Natl Acad Sci U S A. 1987;84:7413–7.
65. Fire A, Xu S, Montgomery MK, Kostas SA, Driver SE, Mello CC. Potent and specific genetic interference by double-stranded RNA in *Caenorhabditis elegans*. Nature. 1998;391:806–11.
66. Doudna JA, Charpentier E. The new frontier of genome engineering with CRISPR-Cas9. Science. 2014;346:1258096.
67. Marx V. Cell-line authentication demystified. Nat Methods. 2014;11:483–8.
68. Nelson-Rees WA. The identification and monitoring of cell line specificity. Prog Clin Biol Res. 1978;26:25–79.
69. Nelson-Rees WA, Flandermeyer RR, Hawthorne PK. Banded marker chromosomes as indicators of intraspecies cellular contamination. Science. 1974;184:1093–6.
70. UKCCCR. UKCCCR guidelines for the use of cell lines in cancer research. Br J Cancer. 2000;82:1495–509.
71. Geraghty RJ, Capes-Davis A, Davis JM, Downward J, Freshney RI, Knezevic I, et al. Cancer Research UK. Guidelines for the use of cell lines in biomedical research. Br J Cancer. 2014;111:1021–46.
72. American Type Culture Collection Standards Development Organization Workgroup ASN-0002. Cell line misidentification: the beginning of the end. Nat Rev Cancer. 2010;10:441–8.
73. Capes-Davis A, Theodosopoulos G, Atkin I, Drexler HG, Kohara A, MacLeod RA, et al. Check your cultures! A list of cross-contaminated or misidentified cell lines. Int J Cancer. 2010;127:1–8.
74. Adey A, Burton JN, Kitzman JO, Hiatt JB, Lewis AP, Martin BK, et al. The haplotype-resolved genome and epigenome of the aneuploid HeLa cancer cell line. Nature. 2013;500:207–11.
75. Nature EDITORIAL. Henrietta lacks: science must right a historical wrong. Nature 2020; 585:7.
76. Wolinetz CD, Collins FS. Recognition of research participants' need for autonomy, remembering the legacy of Henrietta lacks. JAMA. 2020;324:1027–8.

Chapter 2
Establishing Clinical-Grade Standards for Dental Stem Cell Banking: A Pathway to Regenerative Therapy and Personalized Medicine

Mai Mochizuki, Tomoko Kobayashi, and Taka Nakahara

Abstract This chapter explores the establishment of clinical-grade standards for dental stem cell (DSC) banking and its transformative role in regenerative therapy and personalized medicine. DSCs, derived from noninvasively obtained tissues, demonstrate promising proliferative ability, multilineage differentiation potential, and immunomodulatory capacity, positioning them as valuable therapeutic resources. Here, we review the developmental and anatomical origins of DSCs, their regenerative potential, and several clinical successes in tissue repair and regeneration. We detail the rigorous operational and regulatory protocols that ensure the safety and efficacy of DSC-based therapies, focusing on compliance with Japan's Act on the Safety of Regenerative Medicine (ASRM). At the cell processing facility (CPF) of our *Dental Cell Bank*, a specialized platform for cultivating and storing patient-derived (autologous) DSCs, board-certified specialists manage each stage of cell isolation, culture, and cryopreservation, adhering to stringent quality control and standard operating procedures (SOPs). Additionally, we discuss pre-extraction assessments for identifying *cell-bankable* teeth, sterility considerations, and patient guidelines. These procedures are coordinated by certified dental practitioners recognized by our *Dental Cell Bank* and are critical for maintaining cell viability, reducing contamination risks, and optimizing quality. This chapter aims to advance the

M. Mochizuki
Department of Life Science Dentistry, The Nippon Dental University, Chiyoda-ku, Tokyo, Japan

Department of Developmental and Regenerative Dentistry, The Nippon Dental University School of Life Dentistry at Tokyo, Chiyoda-ku, Tokyo, Japan

The Nippon Dental University Dental Cell Bank, Chiyoda-ku, Tokyo, Japan

T. Kobayashi · T. Nakahara (✉)
Department of Developmental and Regenerative Dentistry, The Nippon Dental University School of Life Dentistry at Tokyo, Chiyoda-ku, Tokyo, Japan

The Nippon Dental University Dental Cell Bank, Chiyoda-ku, Tokyo, Japan
e-mail: t.nakahara@tky.ndu.ac.jp

N. Shinomiya, Y. Shimada (eds.), *Applications of Cell Culture*, Current Human Cell Research and Applications, https://doi.org/10.1007/978-981-95-0587-6_2

use of DSCs as a reliable, accessible resource for regenerative therapy, fostering new avenues for clinical application and contributing to the future of personalized medicine.

Keywords Act on the safety of regenerative medicine (ASRM) · Cell processing facility (CPF) · Clinical-grade standards · Clinical trials · Cryopreservation · Dental cell banking · Dental stem cells (DSCs) · Mesenchymal stem/stromal cells (MSCs) · Regenerative medicine · Standard operating procedures (SOPs)

2.1 Introduction

2.1.1 Developmental and Anatomical Origins and Noninvasive Sources of Dental Stem Cells (DSCs)

Dental stem cells (DSCs), a subset of mesenchymal stem/stromal cells (MSCs), are derived from oral tissues that constitute the tooth and its surrounding periodontal structures [1, 2]. These tissues include hard tissues such as enamel, dentin, cementum, and alveolar bone, as well as soft tissues such as dental pulp, periodontal ligament, and gingiva [3]. Additionally, immature teeth with incomplete roots contain the apical papilla, while the surrounding dental soft tissues contain the dental follicle (Fig. 2.1) [4, 5]. With the exception of enamel and gingival epithelium, all of these dental tissues originate in neural crest-derived ectomesenchyme [6–8]. Both permanent and deciduous (primary) teeth provide unique opportunities to observe postnatal developmental processes through imaging techniques such as radiography. This developmental stage is particularly significant in wisdom teeth (third molars), which are frequently extracted due to redundancy and the risk of complications such as misalignment, impaction, and/or infection. Wisdom teeth, whether in the process of development or after root formation, and deciduous teeth, which naturally exfoliate, provide noninvasive sources of DSCs for stem cell research and therapeutic applications [9–12].

2.1.2 Therapeutic Potential of DSCs Through Multilineage Differentiation and Immunomodulation

DSCs exhibit diverse multilineage differentiation potential, allowing them to differentiate not only into dental cells [13, 14] but also into ectodermal [15, 16], mesodermal [17, 18], and even endodermal cells [19–21]. They also possess robust immunomodulatory capabilities, interacting with immune cells and secreting regulatory molecules. This positions them as a promising therapeutic option for treating inflammatory and autoimmune diseases by reducing inflammation and promoting

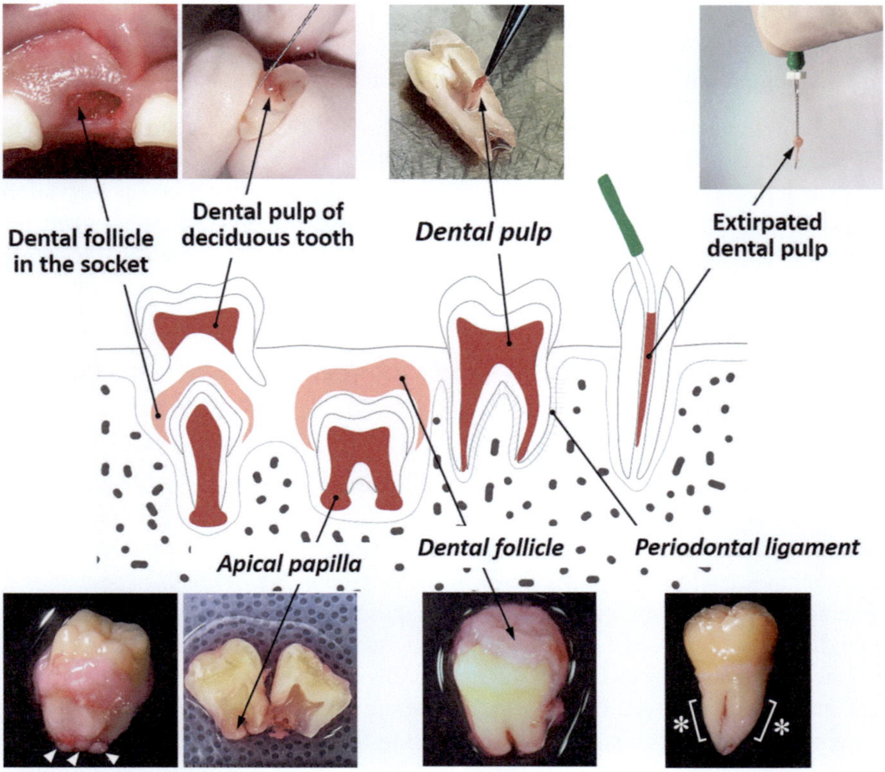

Fig. 2.1 Anatomical and clinical representations of DSC sources. Schematic representation and clinical images of dental structures and tissues used for dental stem cell (DSC) isolation. This includes the dental pulp of permanent and deciduous teeth, the apical papilla (indicated by arrowheads) of immature teeth, and the dental follicle within the extraction socket after deciduous tooth extraction. The periodontal ligament is harvested from the middle third of the completed root (indicated by asterisks)

tissue repair at sites of injury [22–24]. Several types of DSCs, including dental pulp stem cells (DPSCs) [25, 26], stem cells from exfoliated deciduous teeth (SHED) [27, 28], periodontal ligament stem cells (PDLSCs) [29, 30], stem cells from apical papilla (SCAP) [5, 31], and dental follicle stem cells (DFSCs) [32, 33], have been isolated and studied for their potential use in treating a wide range of systemic diseases. Notable clinical trials include the following:

- **DPSCs** have shown success in trials for pulp tissue regeneration, demonstrating the potential to restore functional tissue containing sensory nerves and blood vessels in patients with dental pulp necrosis or irreversible pulpitis [34, 35].
- **SHED** have demonstrated therapeutic efficacy in controlling blood glucose levels in type 2 diabetes mellitus [36] and show potential for promoting functional recovery in acute ischemic stroke [37]. SHED have also proven effective in enhancing orofacial bone regeneration in clinical studies [38, 39].

- **PDLSCs** have been utilized in clinical studies focused on periodontal regeneration, showing success particularly in improving clinical attachment levels and reducing probing depth in patients with chronic periodontitis [40, 41].

These findings align with the growing interest in DSCs for regenerative therapy [42, 43]. Studies have further shown that DSCs can retain their viability, proliferative capacity, and differentiation potential even after long-term cryopreservation [44, 45]. This finding suggests that DSCs can be cryopreserved and banked for future therapeutic use, underscoring their long-term therapeutic potential [46].

2.1.3 Establishing Clinical-Grade Platforms for DSC Banking in Japan

In Japan, where the application of DSCs for cell therapy is advancing, it is essential to establish a safe and reliable platform for supplying DSCs for regenerative therapy [47, 48]. This platform requires the isolation, culture, and cryopreservation of patient-derived (autologous) DSCs from extracted teeth, which necessitates a specialized cell banking system to manage these processes [49]. In response, our team launched the *Dental Cell Bank* in 2015, specifically designed for the culture and storage of autologous DSCs. The stored cells are intended for provision to medical institutions involved in regenerative therapy, thereby facilitating their use in patient-specific treatments. Our cell processing facility (CPF) is the first and only facility among dental universities in Japan to have undergone rigorous review and to have received approval from national regulatory authorities. We strictly adhere to clinical-grade protocols and standard operating procedures (SOPs), ensuring the highest standards of quality and safety throughout cell processing and storage. This positions us as pioneers in Japan, establishing a foundation for the future application of autologous DSCs in regenerative therapy and personalized medicine [50].

This chapter will detail the protocols necessary to maintain clinical-grade standards in the isolation, culture, and cryopreservation of DSCs, ensuring full compliance with the Act on the Safety of Regenerative Medicine (ASRM). It will also outline the certification requirements for CPFs and the operational standards essential for safe and effective clinical use. This includes pre-extraction assessments—such as the selection of *cell-bankable* teeth—and procedures for disinfection, transportation, and processing for cell culture. These protocols support the goal of advancing patient-specific DSC-based therapies.

2.2 Legal Framework and Operational Standards for DSC-Based Therapies

2.2.1 Regulatory Requirements for Regenerative Cell-Based Therapies Under the ASRM

In Japan, the provision of treatments using cells, including DSCs, must strictly comply with the ASRM (Fig. 2.2) [50]. This law is designed to ensure the safety and ethical management of regenerative cell-based therapies. According to the ASRM, institutions must develop a regenerative medicine provision plan that meets the standards established by the Ministry of Health, Labor and Welfare (MHLW). The plan outlines essential details including the cell processing methods, facility safety measures, and the intended therapeutic applications. The ASRM mandates a thorough review and approval by a Certified Special Committee for Regenerative Medicine, an independent body designated by the MHLW, to verify the safety and efficacy of proposed treatments. Once approved, the plan is submitted to the MHLW for final authorization, ensuring compliance with national safety and ethical guidelines. After receiving final authorization, medical institutions are permitted to provide regenerative therapy to patients in accordance with the approved plan.

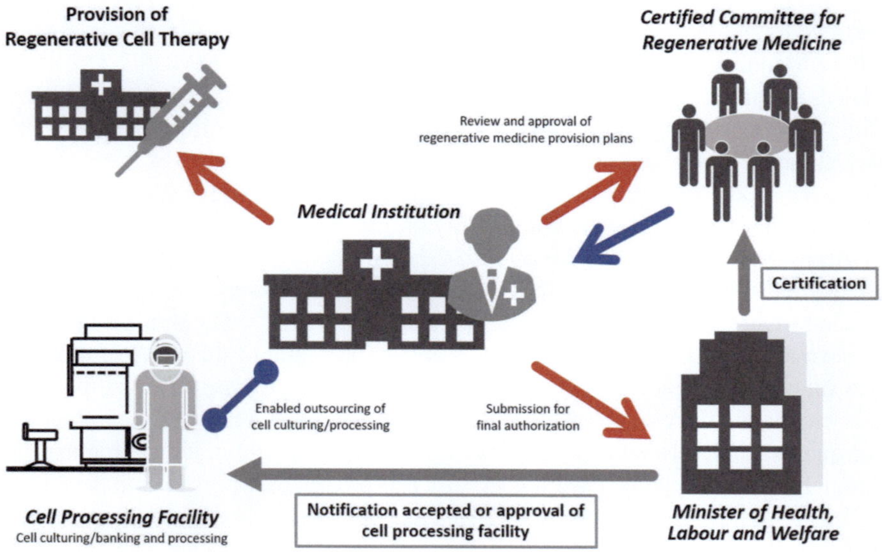

Fig. 2.2 Overview of the regulatory process for regenerative cell therapy in Japan. Flowchart depicting the regulatory process for regenerative cell-based therapies under Japan's Act on the Safety of Regenerative Medicine (ASRM). This includes the roles of certified committees, cell processing facilities (CPFs), the review and approval process by the Ministry of Health, Labour and Welfare, and the provision of regenerative cell therapy by medical institutions. (Adapted from [50] with permission from the Academy of Clinical Dentistry)

2.2.2 Certification and Standards for CPFs

All cell products used in regenerative cell-based therapies, including DSC-based therapies, must be processed at CPFs, also referred to as cell processing centers (CPCs), which require notification to the Regional Bureau of Health and Welfare if within medical institutions, or approval if outside (Fig. 2.2). This certification process ensures that facilities meet stringent standards for quality, sterility, and safety, minimizing risks to patients. Certification must be renewed every 5 years to ensure ongoing adherence to regulatory requirements. SOPs play a critical role in maintaining consistency and safety, ensuring that each step of cell processing is controlled and documented. This regulatory framework enhances the credibility of regenerative therapy in Japan and ensures that patients receive treatments adhering to the highest standards of quality and safety.

2.2.3 Clinical and Operational Standards at Nippon Dental University's Dental Cell Bank

At the *Dental Cell Bank* of the Nippon Dental University (NDU), a specialized platform for cultivating and storing patient's own DSCs, all processes are conducted within a highly robust CPF [50]. The facility is operated by a team of professionals certified by the Japanese Society for Regenerative Medicine (JSRM), including one JSRM-Manufacturing Manager in Cell Processing Facility (J-MMCPF: T.N.), two JSRM-Advanced Cell Processing Operators (J-ACPO: M.M., T.K.), one JSRM-Cell Processing Operator (J-CPO), and two JSRM-Certified Dentists for Regenerative Medicine (J-CDRM: T.N., M.M.). This well-structured team ensures that every stage of cell processing and storage complies with rigorous SOPs to maintain the sterility, quality, and viability of DSCs.

An essential factor for successful DSC culture is the condition of the extracted teeth and the patient's oral health. Infection and inflammation of the teeth and periodontal tissues can significantly affect the viability of DSCs. Therefore, a thorough pre-extraction assessment is critical to evaluate the health of the entire oral cavity, focusing on the target teeth and periodontal tissues, and ensuring the sterility of the extracted teeth, thereby optimizing the conditions for cell culture. To maintain tissue quality for cell isolation, the extraction procedure should minimize trauma and thermal stimulation while ensuring sterility and prompt transportation to the CPF.

To uphold these standards, our cell bank has implemented a certification program for dentists, physicians, dental hygienists, and dental technicians, ensuring that all procedures are performed by qualified professionals [50]. To date, more than 1350 professionals have been certified, reflecting the growing adoption of high-quality standards in dental cell banking. Within this framework, certified dentists are responsible for educating patients on dental cell banking, obtaining informed consent, and performing necessary tooth extractions. Certified physicians play a key role in informing patients about regenerative therapy and referring those interested in cell banking to certified dentists. Certified coordinators, consisting of dental hygienists and dental technicians, manage patient interactions and process extracted teeth, supporting certified dentists in fulfilling their duties effectively. This overview provides a comprehensive understanding of the clinical-grade manufacturing processes essential for therapeutic DSC applications. The following sections will first outline the clinical precautions taken during the tooth extraction process to optimize stem cell quality and viability, followed by detailed procedures for the isolation, culture, and cryopreservation of DSCs at our CPF.

2.3 Selection Criteria for Cell-Bankable Teeth to Ensure Sterility

Given that cells transplanted in cell therapy are viable, it is technically challenging to achieve complete sterility of the final cell product. Therefore, strict aseptic techniques throughout the manufacturing process ensure the sterility of the final cell product. It is essential to implement quality management strategies at each stage of the manufacturing process, from the environment and source materials to the processing materials and procedures, to minimize the risk of microbial contamination. When extracted teeth are used as source materials for cell banking, establishing suitability criteria for these teeth is a crucial step. Due to the high microbial load in the oral cavity, which makes sterilization of extracted teeth impossible, it is essential to appropriately select teeth suitable for cell banking (i.e., *cell-bankable* teeth) before extraction and to maintain optimal oral hygiene during extraction. Furthermore, ensuring safety and temperature control during transportation of extracted teeth is imperative.

To obtain teeth for cell banking, patients who provide informed consent for the procedure undergo curative tooth extraction by certified dentists recognized by our cell bank. Extractions are performed by the certified dentists at dental clinics nationwide, as well as at the *Dental Cell Bank Clinics* at the NDU Hospitals in Tokyo and Niigata. If the suitability of a tooth cannot be definitively determined based on clinical findings, it is recommended to err on the side of caution by deeming the tooth unsuitable to ensure sterility. The clinical criteria for suitable teeth for cell banking are as follows:

2.3.1 Teeth Requiring Curative Extraction

As a fundamental criterion, teeth suitable for cell banking are those for which curative tooth extraction is clinically indicated; in other words, we do not recommend removing healthy teeth solely for cell banking purposes. According to this

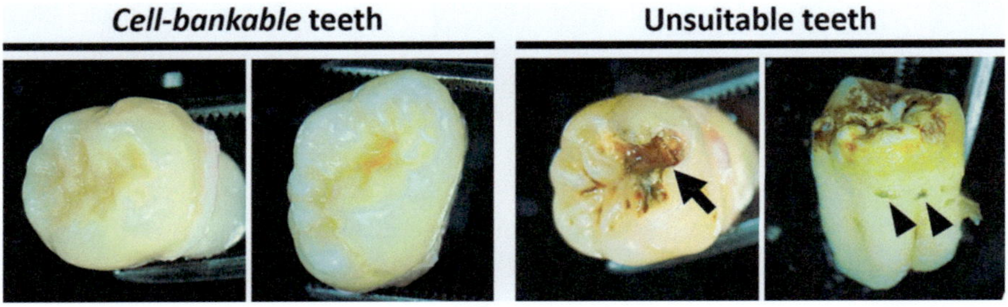

Fig. 2.3 Selection criteria for cell-bankable and unsuitable teeth for DSC banking. Photographic comparison showing examples of teeth considered suitable (i.e., *cell-bankable*) and unsuitable for dental stem cell (DSC) banking. Unsuitable teeth show dental caries (arrow) and calculus (arrowheads), which compromise sterility and cell viability

guideline, appropriate teeth include naturally exfoliating primary teeth. Impacted or nonfunctional third molars are also considered eligible. Teeth (e.g., premolars) extracted for orthodontic purposes to create space or improve alignment are likewise considered eligible. Additionally, supernumerary teeth—extra teeth that have developed beyond the normal dental formula—are suitable for cell banking when they require removal to prevent crowding and are structurally intact and uninfected. These criteria ensure that only non-infected, structurally intact teeth are selected for cell banking, optimizing the quality and viability of DSCs for therapeutic applications.

2.3.2 Teeth Without Caries or Periodontal Disease (Healthy Teeth)

As shown in Fig. 2.3, the target teeth must be free of caries (Fig. 2.3; arrow), dental plaque, and calculus (Fig. 2.3; arrowheads). However, because periodontal pathogens are not visible, certified practitioners must determine their suitability based on patient symptoms, clinical examination, and diagnosis.

2.3.3 Teeth Without Dental Treatment (Healthy Teeth)

The teeth must not have undergone any past dental crown restoration, prosthetics, or preventive sealing (dental sealant). Even in the absence of symptoms, such teeth might be infected by bacteria through gaps in restorations, prosthetics, or sealants. These contaminants cannot be removed, rendering the teeth unsuitable for cell banking. Additionally, restored teeth are challenging to assess through radiographic examination, and ensuring sterility is not feasible, necessitating their exclusion.

2.3.4 *Pulp Tissue Extracted for Treating Dentin Hypersensitivity*

Although the tooth is healthy and free of caries or periodontal disease, in cases where the patient cannot endure pain caused by dentin hypersensitivity, triggered by exposure to hot or cold stimuli, pulp extirpation may be performed as a necessary treatment (Fig. 2.1). The removed pulp tissue is free of bacterial infection and remains viable, making it suitable.

2.3.5 *Pulp Tissue Obtained from Elective Pulp Extirpation*

Even if the tooth is healthy and free of caries or periodontal disease, pulp tissue may be obtained from elective pulp extirpation during prosthetic treatment (Fig. 2.1). Similar to pulp extirpation for dentin hypersensitivity, pulp tissue can be aseptically isolated and used for cell culture.

2.4 Patient Guidance and Oral Environment Improvement

Once a tooth is determined to be suitable by certified dentists, it is necessary to improve the oral environment to prevent microbial infection of the harvested tissue prior to extraction. Complete asepsis is not achievable due to the presence of oral bacteria. However, with appropriate patient instructions, including dietary restrictions and oral hygiene prior to extraction, the risk of microbial contamination of harvested tissue for cell banking can be minimized.

2.4.1 *Dietary Restrictions Prior to Extraction: Avoiding Natto and Dairy Products*

Patients should be advised to avoid consuming fermented foods such as natto and dairy products on the day of the procedure. *Bacillus subtilis* in natto and lactic acid bacteria in dairy products are difficult to eliminate through standard oral cleaning; therefore, these foods should be avoided on the day of extraction.

2.4.2 Preparation of the Oral Environment for Tooth Extraction and Pulp Extirpation

Prepare the oral environment for tooth extraction and pulp extirpation. Specifically, perform professional mechanical tooth cleaning (PMTC) by a certified dentist or certified coordinator (dental hygienist) before extraction, and perform scaling (removal of dental calculus) if necessary. Additionally, PMTC should be repeated before the procedure, and patients should be instructed to use mouthwash. It is essential to improve the cleanliness of the entire oral cavity, not just the target tooth.

2.5 Disinfection of Extracted Teeth

Harvested tissues, including extracted teeth and pulp, must be safely transported to the CPF while maintaining a high degree of cleanliness. To achieve this, appropriate processing of source materials must be conducted immediately after extraction.

2.5.1 Disinfection and Storage of Extracted Teeth

After extraction, teeth should be disinfected and rinsed chairside and then placed in a preservation solution for transport. Specifically, a 100 mL disinfection tube containing Hanks' balanced salt solution (HBSS; Fujifilm Wako Pure Chemical, Tokyo, Japan) with 0.002% povidone-iodine (Shionogi Healthcare, Osaka, Japan) and 0.03% egg albumin (Nacalai Tesque, Kyoto, Japan) is shaken to create foam for thorough disinfection. After foam disinfection, the tooth is rinsed with HBSS and placed in a preservation solution, Dulbecco's Modified Eagle Medium/Nutrient Mixture F-12 (DMEM/F12; Fujifilm Wako Pure Chemical) containing 50 U/mL penicillin and 50 μg/mL streptomycin (P/S; Nacalai Tesque). The tooth is then stored at 4 °C until pulp tissue is collected at the CPF.

2.5.2 Handling of Extracted Pulp and Residual Dental Soft Tissues

Pulp tissue obtained through pulp extirpation, as well as residual dental soft tissues in extraction sockets (such as the residual pulp from deciduous teeth or the dental follicle of succeeding permanent teeth) (Fig. 2.1), is less susceptible to contamination by saliva than extracted teeth. Therefore, these tissues can be placed directly in a 2 mL microtube containing preservation solution without the need for disinfection and rinsing.

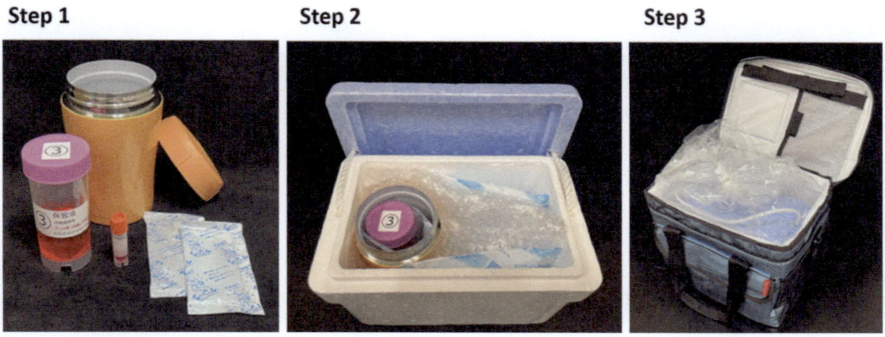

Fig. 2.4 Transportation method for harvested dental tissues. Photographic step-by-step depiction of the process for transporting extracted teeth and/or dental soft tissues in preservation solution. The process involves placing the sample tubes in an insulated container with ice packs (Step 1), securing in a Styrofoam container with additional ice packs (Step 2), and sealing in a plastic bag with a cable tie before placing in an insulated bag for refrigerated courier transport (Step 3). Transport is maintained at approximately 4 °C, with culture initiation within 72 h

2.6 Transportation of Harvested Tissues

Extracted teeth and/or dental soft tissues in preservation solution are transported under refrigeration at approximately 4 °C. Specifically, a 100 mL preservation tube containing extracted tooth and/or a 2 mL microtube containing dental soft tissue is placed in an insulated container (thermal pot) with ice packs (Fig. 2.4; Step 1), which is then placed in a Styrofoam container with additional ice packs (Fig. 2.4; Step 2). The Styrofoam container is placed in a plastic bag sealed with a cable tie and then placed into an insulated bag for transportation by a refrigerated courier (Fig. 2.4; Step 3). The time from extraction to initiation of culture is generally within 72 h.

2.7 Processing of Extracted Teeth and Cell Culture

2.7.1 Preparation at CPF

Extracted teeth serve as the source material for cell isolation; however, they may also introduce contamination during cell culture. Therefore, dedicated safety cabinets are used for processing extracted teeth and are distinct from those used for cell culture at the CPF. For permanent teeth, such as wisdom teeth and premolars, the teeth must be split to extract pulp tissue from the pulp chamber. The safety cabinet designated for splitting extracted teeth is equipped with a vacuum device to collect debris generated when grinding teeth using dental cutting burs. Given that this procedure generates heat that may lead to cell death within the tooth, cooling with HBSS is also essential. In contrast, in the case of deciduous teeth, root resorption is typically present, exposing the pulp chamber. In immature permanent teeth, often impacted, the apical papilla at the apex of the root is also exposed, facilitating the easy extraction of pulp tissue without splitting the tooth. Deciduous and impacted teeth are often surrounded by dental soft tissue, such as dental follicle, which can be processed similarly to pulp tissue. Under these preparatory conditions, extracted teeth delivered to the CPF are processed to allow for cell isolation and culture.

2.7.2 *Dental Tissue Extraction and Cell Isolation (Primary Culture)*

1. In the safety cabinet designated for tooth processing, add 0.002% povidone-iodine to the preservation tube containing an extracted tooth and rinse it.
2. Place the extracted tooth in a 100 mm culture dish with HBSS + P/S and make incision marks using a dental cutting bur while cooling with HBSS.
3. Rinse the extracted tooth with HBSS.
4. Transfer it to a 60 mm culture dish with HBSS and move it to another safety cabinet for cell culture.
5. Split the tooth using an elevator (Heber) and extract the pulp tissue (Fig. 2.5).
6. Rinse the pulp tissue with HBSS + P/S.
7. In a separate 60 mm culture dish, place 100 μL of xeno-/serum-free culture medium (MSC Expansion XSFM B; Irvine Scientific, Santa Ana, CA, USA) supplemented with P/S, and finely chop the pulp tissue into fragments smaller than 1 mm^3 using a scalpel. For immature tooth, the apical papilla at the root apex is separated from the pulp within the pulp chamber (Fig. 2.1; arrowheads), and each tissue is chopped separately. If dental follicle tissue is attached (Fig. 2.1), it is similarly chopped in another culture dish. In mature teeth, where periodontal ligament (PDL) tissue is to be harvested, to avoid contamination with other cell types, the PDL tissue is shaved off from the middle third of the

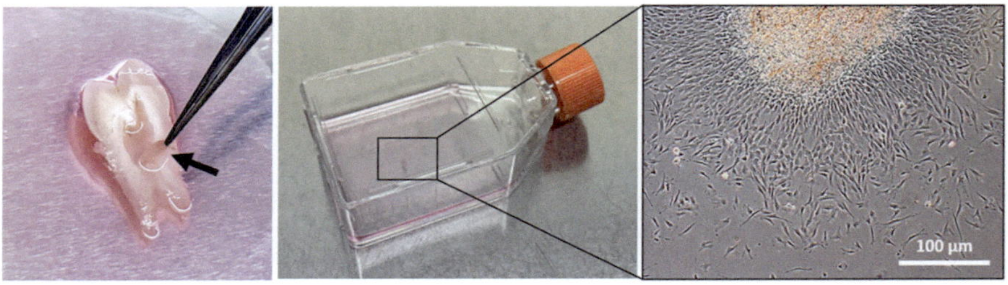

Fig. 2.5 Procedure for extracting and processing pulp tissue. Photographic depiction of extraction and cell culture processes for dental pulp. This includes the removal of pulp tissue using micro-forceps after initial tooth disinfection and splitting, followed by culturing the finely chopped pulp tissue in a cell culture flask

root length using a scalpel (Fig. 2.1; asterisks) and finely chopped in the same manner as above.

8. Transfer the chopped tissue into a 25 cm^2 culture flask (CellBIND® surface; Corning, NY, USA) containing 1 mL of culture medium, and culture it in an incubator at 37 °C with 5% CO_2 (Fig. 2.5).
9. Replace the medium every 3–4 days, and continue culturing until approximately 90% confluence is reached.
10. For enzymatic isolation of cells from harvested tissue, a mixture of 0.8 PZ-U/mL collagenase (SERVA Electrophoresis, Heidelberg, Germany) and 1200 PU/mL dispase (Sanko Junyaku, Tokyo, Japan) is used for digestion at 37 °C for 60 min. The digested solution is filtered through a 70 μm cell strainer to obtain a single-cell suspension, and is seeded into a culture flask [51].

2.7.3 *Subculture*

1. Observe the cultured cells under a phase-contrast microscope.
2. Confirm approximately 90% confluence, and collect the culture supernatant in a 50 mL tube.
3. Wash the cells with HBSS, then add 1 mL of TrypLE™ Express Enzyme (Thermo Fisher Scientific, Waltham, MA, USA).
4. After 1–3 min, gently agitate the culture flask and confirm cell detachment under the microscope.
5. Add 3 mL of HBSS to the culture flask and collect the detached cells into the 50 mL tube.
6. Count the cells in the suspension using a Countess™ automatic cell counter (Thermo Fisher Scientific).
7. Adjust the volume of the cell suspension to 40 mL with HBSS, then pipette gently.
8. Centrifuge at 120 × *g* for 5 min at room temperature.
9. Discard the supernatant and add the appropriate culture medium.
10. Seed the cells at a density of 3×10^5 cells per 25 cm^2 in a culture flask.
11. Culture the cells in an incubator at 37 °C with 5% CO_2.
12. Replace the medium every 3–4 days, continuing until approximately 90% confluence is reached.

2.7.4 *Cryopreservation*

1. Observe the cultured cells under a phase-contrast microscope.
2. Confirm approximately 90% confluence, and collect the culture supernatant in a 50 mL tube.

3. Perform mycoplasma testing (MycoAlert™ Mycoplasma Detection Kit; Lonza, Verviers, Belgium) using a portion of the culture supernatant.
4. Wash the cells with HBSS, and add 1 mL of TrypLE™ Express Enzyme.
5. After 1–3 min, gently agitate the culture flask and confirm cell detachment under the microscope.
6. Add 3 mL of HBSS to the culture flask and collect the detached cells into the 50 mL tube.
7. Count the cells in the suspension using a Countess™ automatic cell counter.
8. Adjust the volume of the cell suspension to 40 mL with HBSS, and pipette gently.
9. Centrifuge at 120 × *g* for 5 min at room temperature.
10. Discard the supernatant and add STEM-CELLBANKER® DMSO free GMP grade (Takara Bio, Shiga, Japan) to achieve a concentration of 1.0×10^6 cells/mL.
11. Transfer the cell suspension into cryotubes, each containing approximately 1.0×10^6 cells.
12. Place the cryotubes containing cells into a CoolCell® container (Corning) and store at −80 °C.
13. Within a few days, transfer the cryotubes to a liquid nitrogen storage container.

2.7.5 Considerations for Protocol Compliance and Safety

The culture protocol described above is continuously reviewed and modified to improve safety and efficacy, addressing any advancements and challenges observed during its practical application. It is recommended that the culture media and materials used for cell culture in this protocol obtain certifications confirming their suitability for regenerative therapy, such as the "Certificate of Material Suitability for Regenerative Medicine Products" issued by Japan's Pharmaceuticals and Medical Devices Agency (PMDA). Furthermore, all biological materials and reagents used in culture must comply with the "Standards for Biological Raw Materials" to ensure that they are safe, of high quality, and free from contaminants. Adherence to these standards and obtaining the necessary certifications help minimize the risks related to biological contamination, ensuring the reliability of the final cell products.

2.8 Characterization of DSCs

DSCs exhibit cellular characteristics that align with the criteria established by the International Society for Cell and Gene Therapy (ISCT) for MSCs [52]. According to ISCT guidelines, MSCs must meet specific standards, including adherence to plastic under standard culture conditions; expression of surface markers such as CD105, CD73, and CD90; and lack of expression of hematopoietic markers such as CD45, CD34, CD14 or CD11b, CD79α or CD19, and HLA-DR. Additionally,

MSCs must possess the capacity for trilineage differentiation into osteoblasts, adipocytes, and chondroblasts under standard in vitro conditions.

In addition to meeting these MSC criteria, we have established our own criteria for DSCs based on our previous reports [5, 30, 33, 51, 53]. Specifically, for cells that have undergone three or more passages after expansion culture, we evaluate proliferative capacity, surface antigen expression, and multilineage differentiation potential. We define DSCs with MSC characteristics as those that meet the following three criteria: (1) sufficient proliferative capacity, (2) expression of appropriate MSC surface markers, and (3) confirmed potential for multilineage differentiation. Moreover, for DPSCs, we consider the unique ability to form *self-multilayered cell sheets* to be a desirable characteristic (patent pending) [53], with potential applications in cell sheet therapy for systemic diseases. This property is particularly significant, as it supports the development of advanced therapeutic approaches using autologous DPSC-derived cell sheets [11, 14, 54].

2.9 Conclusions

The establishment of clinical-grade standards for DSC banking is a significant step forward in cell-based therapies. This chapter outlined the developmental and anatomical origins and therapeutic potential of DSCs, along with protocols for their isolation, culture, and cryopreservation, thereby demonstrating their feasibility for clinical use. By adhering to rigorous standards and regulatory compliance, the *Dental Cell Bank* ensures the consistent delivery of high-quality autologous DSCs. Comprehensive assessments—including the selection of *cell-bankable* teeth—and patient guidance support the success of DSC banking. Continued refinement of these protocols and compliance with evolving guidelines will be key to maximizing their benefits. DSCs are poised to shape the future of regenerative therapy and personalized medicine, opening new avenues for treating complex conditions and improving patient care.

Acknowledgments This work was supported in part by a Grant-in-Aid for Young Scientists (No. 23K16050 to M.M.), a Grant-in-Aid for Scientific Research (C) (No. 23K09365 to T.K., M.M., T.N.), a Grant-in-Aid for Challenging Research (Exploratory) (No. 23K18367 to T.N., M.M., T.K.), and a Grant-in-Aid for Scientific Research (B) (No. 21H03147 to T.N., M.M., T.K.), all provided by the Japan Society for the Promotion of Science (JSPS KAKENHI).

References

1. Bakopoulou A, About I. Stem cells of dental origin: current research trends and key milestones towards clinical application. Stem Cells Int. 2016;2016:4209891.
2. Huang GT, Gronthos S, Shi S. Mesenchymal stem cells derived from dental tissues vs. those from other sources: their biology and role in regenerative medicine. J Dent Res. 2009;88:792–806.

3. Nakahara T, Ide Y. Tooth regeneration: implications for the use of bioengineered organs in first-wave organ replacement. Hum Cell. 2007;20:63–70.
4. Odontology prize 2017. Odontology. 2017;105:391.
5. Tamaki Y, Nakahara T, Ishikawa H, Sato S. In vitro analysis of mesenchymal stem cells derived from human teeth and bone marrow. Odontology. 2013;101:121–32.
6. Al Madhoun A, Sindhu S, Haddad D, Atari M, Ahmad R, Al-Mulla F. Dental pulp stem cells derived from adult human third molar tooth: a brief review. Front Cell Dev Biol. 2021;9:717624.
7. Mayo V, Sawatari Y, Huang CY, Garcia-Godoy F. Neural crest-derived dental stem cells—where we are and where we are going. J Dent. 2014;42:1043–51.
8. Chai Y, Jiang X, Ito Y, Bringas P Jr, Han J, Rowitch DH, et al. Fate of the mammalian cranial neural crest during tooth and mandibular morphogenesis. Development. 2000;127:1671–9.
9. Shi X, Mao J, Liu Y. Pulp stem cells derived from human permanent and deciduous teeth: biological characteristics and therapeutic applications. Stem Cells Transl Med. 2020;9:445–64.
10. Gan L, Liu Y, Cui D, Pan Y, Zheng L, Wan M. Dental tissue-derived human mesenchymal stem cells and their potential in therapeutic application. Stem Cells Int. 2020;2020:8864572.
11. Nakahara T. A review of new developments in tissue engineering therapy for periodontitis. Dent Clin N Am. 2006;50:265–76, ix–x.
12. Rahaman MN, Mao JJ. Stem cell-based composite tissue constructs for regenerative medicine. Biotechnol Bioeng. 2005;91:261–84.
13. El-Rashidy AA, Radwan IA, Rady D, El Moshy S, Abbass MMS, Sadek KM, et al. Dental mesenchymal stem/progenitor cells: a new prospect in regenerative medicine. Stem Cells. 2021:135–56.
14. Nakahara T. Potential feasibility of dental stem cells for regenerative therapies: stem cell transplantation and whole-tooth engineering. Odontology. 2011;99:105–11.
15. Arimura Y, Shindo Y, Yamanaka R, Mochizuki M, Hotta K, Nakahara T, et al. Peripheral-neuron-like properties of differentiated human dental pulp stem cells (hDPSCs). PLoS One. 2021;16:e0251356.
16. Xiao L, Ide R, Saiki C, Kumazawa Y, Okamura H. Human dental pulp cells differentiate toward neuronal cells and promote neuroregeneration in adult organotypic hippocampal slices in vitro. Int J Mol Sci. 2017;18:1745.
17. Kobayashi T, Torii D, Iwata T, Izumi Y, Nasu M, Tsutsui TW. Characterization of proliferation, differentiation potential, and gene expression among clonal cultures of human dental pulp cells. Hum Cell. 2020;33:490–501.
18. Kerkis I, Kerkis A, Dozortsev D, Stukart-Parsons GC, Gomes Massironi SM, Pereira LV, et al. Isolation and characterization of a population of immature dental pulp stem cells expressing OCT-4 and other embryonic stem cell markers. Cells Tissues Organs. 2006;184:105–16.
19. Aly RM, Aglan HA, Eldeen GN, Mahmoud NS, Aboul-Ezz EH, Ahmed HH. Efficient generation of functional pancreatic beta cells from dental-derived stem cells via laminin-induced differentiation. J Genet Eng Biotechnol. 2022;20:85.
20. Iwanaka T, Yamaza T, Sonoda S, Yoshimaru K, Matsuura T, Yamaza H, et al. A model study for the manufacture and validation of clinical-grade deciduous dental pulp stem cells for chronic liver fibrosis treatment. Stem Cell Res Ther. 2020;11:134.
21. Yokoyama T, Yagi Mendoza H, Tanaka T, Ii H, Takano R, Yaegaki K, et al. Regulation of CCl(4)-induced liver cirrhosis by hepatically differentiated human dental pulp stem cells. Hum Cell. 2019;32:125–40.
22. Li P, Ou Q, Shi S, Shao C. Immunomodulatory properties of mesenchymal stem cells/dental stem cells and their therapeutic applications. Cell Mol Immunol. 2023;20:558–69.
23. Shang L, Shao J, Ge S. Immunomodulatory functions of oral mesenchymal stem cells: novel force for tissue regeneration and disease therapy. J Leukoc Biol. 2021;110:539–52.
24. Paganelli A, Trubiani O, Diomede F, Pisciotta A, Paganelli R. Immunomodulating profile of dental mesenchymal stromal cells: a comprehensive overview. Front Oral Health. 2021;2:635055.

25. Gancheva MR, Kremer KL, Gronthos S, Koblar SA. Using dental pulp stem cells for stroke therapy. Front Neurol. 2019;10:422.
26. Gronthos S, Mankani M, Brahim J, Robey PG, Shi S. Postnatal human dental pulp stem cells (DPSCs) in vitro and in vivo. Proc Natl Acad Sci U S A. 2000;97:13625–30.
27. Yadav P, Vats R, Bano A, Namdev R, Bhardwaj R. Ameliorative potential of stem cells from human exfoliated deciduous teeth (SHED) in preclinical studies: a meta-analysis. Regen Ther. 2023;24:117–34.
28. Miura M, Gronthos S, Zhao M, Lu B, Fisher LW, Robey PG, et al. SHED: stem cells from human exfoliated deciduous teeth. Proc Natl Acad Sci U S A. 2003;100:5807–12.
29. Seo BM, Miura M, Gronthos S, Bartold PM, Batouli S, Brahim J, et al. Investigation of multipotent postnatal stem cells from human periodontal ligament. Lancet. 2004;364:149–55.
30. Murabayashi D, Mochizuki M, Tamaki Y, Nakahara T. Practical methods for handling human periodontal ligament stem cells in serum-free and serum-containing culture conditions under hypoxia: implications for regenerative medicine. Hum Cell. 2017;30:169–80.
31. Sonoyama W, Liu Y, Yamaza T, Tuan RS, Wang S, Shi S, et al. Characterization of the apical papilla and its residing stem cells from human immature permanent teeth: a pilot study. J Endodont. 2008;34:166–71.
32. Morsczeck C, Gotz W, Schierholz J, Zeilhofer F, Kuhn U, Mohl C, et al. Isolation of precursor cells (PCs) from human dental follicle of wisdom teeth. Matrix Biol. 2005;24:155–65.
33. Shinagawa-Ohama R, Mochizuki M, Tamaki Y, Suda N, Nakahara T. Heterogeneous human periodontal ligament-committed progenitor and stem cell populations exhibit a unique cementogenic property under in vitro and in vivo conditions. Stem Cells Dev. 2017;26:632–45.
34. Xuan K, Li B, Guo H, Sun W, Kou X, He X, et al. Deciduous autologous tooth stem cells regenerate dental pulp after implantation into injured teeth. Sci Transl Med. 2018;10:eaaf3227.
35. Nakashima M, Fukuyama F, Iohara K. Pulp regenerative cell therapy for mature molars: a report of 2 cases. J Endodont. 2022;48:1334–40.e1.
36. Li W, Jiao X, Song J, Sui B, Guo Z, Zhao Y, et al. Therapeutic potential of stem cells from human exfoliated deciduous teeth infusion into patients with type 2 diabetes depends on basal lipid levels and islet function. Stem Cells Transl Med. 2021;10:956–67.
37. Suda S, Nito C, Ihara M, Iguchi Y, Urabe T, Matsumaru Y, et al. Randomised placebo-controlled multicentre trial to evaluate the efficacy and safety of JTR-161, allogeneic human dental pulp stem cells, in patients with Acute Ischaemic stRoke (J-REPAIR). BMJ Open. 2022;12:e054269.
38. Hernandez-Monjaraz B, Santiago-Osorio E, Ledesma-Martinez E, Alcauter-Zavala A, Mendoza-Nunez VM. Retrieval of a periodontally compromised tooth by allogeneic grafting of mesenchymal stem cells from dental pulp: a case report. J Int Med Res. 2018;46:2983–93.
39. Tanikawa DYS, Pinheiro CCG, Almeida MCA, Oliveira C, Coudry RA, Rocha DL, et al. Deciduous dental pulp stem cells for maxillary alveolar reconstruction in cleft lip and palate patients. Stem Cells Int. 2020;2020:6234167.
40. Iwata T, Yamato M, Washio K, Yoshida T, Tsumanuma Y, Yamada A, et al. Periodontal regeneration with autologous periodontal ligament-derived cell sheets—a safety and efficacy study in ten patients. Regen Ther. 2018;9:38–44.
41. Sanchez N, Fierravanti L, Nunez J, Vignoletti F, Gonzalez-Zamora M, Santamaria S, et al. Periodontal regeneration using a xenogeneic bone substitute seeded with autologous periodontal ligament-derived mesenchymal stem cells: a 12-month quasi-randomized controlled pilot clinical trial. J Clin Periodontol. 2020;47:1391–402.
42. Casagrande L, Cordeiro MM, Nör SA, Nör JE. Dental pulp stem cells in regenerative dentistry. Odontology. 2011;99:1–7.
43. Song WP, Jin LY, Zhu MD, Wang H, Xia DS. Clinical trials using dental stem cells: 2022 update. World J Stem Cells. 2023;15:31–51.
44. Paes SM, Pupo YM, Cavenago BC, Fonseca-Silva T, Santos CCO. Cryopreservation of mesenchymal stem cells derived from dental pulp: a systematic review. Restor Dent Endod. 2021;46:e26.

45. Brozek R, Dorocka-Bobkowska B, Kurpisz M. Long-term cryopreservation of dental stem cells. J Physiol Pharmacol. 2023;74.
46. Wang X, Li F, Wu S, Xing W, Fu J, Wang R, et al. Research progress on optimization of in vitro isolation, cultivation and preservation methods of dental pulp stem cells for clinical application. Front Bioeng Biotechnol. 2024;12:1305614.
47. Ohkoshi S, Hirono H, Nakahara T, Ishikawa H. Dental pulp cell bank as a possible future source of individual hepatocytes. World J Hepatol. 2018;10:702–7.
48. Honda M, Ohshima H. Biological characteristics of dental pulp stem cells and their potential use in regenerative medicine. J Oral Biosci. 2022;64:26–36.
49. Nakahara T. Patient-derived stem cell medicine: a safe and reliable cell processing and regenerative therapeutic application of human dental pulp stem cells. In: Morimoto Y, Nakahara T, editors. Cell processing technology. Singapore: Springer; 2024. p. 105–28.
50. Mochizuki M, Nakahara T. Three approaches to overcome the disadvantages of regenerative medicine in dentistry: making regenerative medicine an "accessible" medical treatment. J Acad Clin Dent. 2023;42:323–30. (in Japanese).
51. Mochizuki M, Nakahara T. Establishment of xenogeneic serum-free culture methods for handling human dental pulp stem cells using clinically oriented in-vitro and in-vivo conditions. Stem Cell Res Ther. 2018;9:25.
52. Dominici M, Le Blanc K, Mueller I, Slaper-Cortenbach I, Marini F, Krause D, et al. Minimal criteria for defining multipotent mesenchymal stromal cells. The International Society for Cellular Therapy position statement. Cytotherapy. 2006;8:315–7.
53. Mochizuki M, Sagara H, Nakahara T. Type I collagen facilitates safe and reliable expansion of human dental pulp stem cells in xenogeneic serum-free culture. Stem Cell Res Ther. 2020;11:267.
54. Nakahara T. Tooth/periodontal organ engineering: the ultimate goal of tooth regeneration. In: Greco G, editor. Tissue engineering research trends. New York: Nova Science Publishers; 2008. p. 7–13.

Chapter 3
Maintaining the Genotype and Phenotype of Human Cell Lines Through Optimization of Culture Conditions, Passaging, and Differentiation Induction

Aiki Marushima and **Hiroshi Ishikawa**

Abstract The successful establishment and long-term maintenance of human cell lines require strict control of the culture conditions to ensure phenotypic stability, genomic integrity, and experimental reproducibility. This review provides practical guidelines for culturing established and newly derived cell lines, focusing on three core factors: substrate coating, culture media composition, and environmental consistency. Appropriate coating materials are essential for supporting cell adhesion and lineage-specific function. Culture media must be carefully selected and supplemented to match the metabolic and differentiation needs of the target cell type. Additionally, environmental parameters such as oxygen concentration should be optimized. Consistent passaging protocols and enzymatic dissociation techniques as well as regular authentication are also essential for maintaining cell identity and function. Together, these strategies form the foundation for reliable in vitro cell culture, enabling applications in disease modeling and regenerative medicine.

Keywords Cell lines · Culture condition · Passaging protocol · Differentiation induction

3.1 Introduction

During the culture and passage of established cell lines, it is not uncommon for cellular characteristics to drift or for genomic abnormalities to accumulate. To minimize these risks, it is critical to optimize culture conditions and protocols for the

A. Marushima (✉) · H. Ishikawa
Laboratory of Clinical Regenerative Medicine, Department of Neurosurgery, Institute of Medicine, University of Tsukuba, Tsukuba, Ibaraki, Japan
e-mail: aiki.marushima@md.tsukuba.ac.jp

N. Shinomiya, Y. Shimada (eds.), *Applications of Cell Culture*, Current Human Cell Research and Applications, https://doi.org/10.1007/978-981-95-0587-6_3

specific cell type, including media composition, oxygen tension, passaging frequency and method, cell selection strategies, and monitoring techniques.

In this section, we describe practical methodologies for the **maintenance, passaging, and differentiation induction** of established cell lines based on protocols developed and implemented in our laboratory [1–13]. Particular attention is given to stem cell-derived lines, strategies for directed differentiation toward target cell types, and approaches for evaluating the phenotypic and functional properties of differentiated cells. These practices are intended to ensure the reproducibility, stability, and translational relevance of cultured cells for both basic research and therapeutic applications.

3.2 Substrate Coatings for Established Cell Lines

Epithelial cells (ECs) in situ maintain contact with a basement membrane, so the development of substrate coatings for ECs began with attempts to replicate the basement membrane composition. The appropriate culture surface coating is a critical step for the establishment and subsequent maintenance of adherent cell lines, especially those with relatively weak innate attachment properties, because adhesion influences both proliferation rate and differentiation status. Thus, the choice of coating materials and methods must be tailored both to the specific cell type and the intended application.

Cell lines derived from malignant tumors often proliferate rapidly on standard tissue culture plastic without additional coating. However, cells derived from normal tissues, especially naturally immortalized cell lines, frequently require coating with native (physiologically innate) matrix components to ensure proper adhesion and phenotype preservation. In general, the same coating method used during cell line establishment should be applied for routine maintenance [14, 15].

3.2.1 Common Coating Materials and Their Applications

Poly-l-lysine: Enhances the adhesion of weakly adherent cells such as neurons and endothelial cells. Frequently used in primary cultures and neural induction protocols.

Collagen: A primary component of the extracellular matrix (ECM), collagen supports the adhesion and growth of muscle cells, fibroblasts, epithelial cells, and endothelial cells by providing a physiologically relevant microenvironment.

Fibronectin: An ECM glycoprotein that promotes the adhesion and spread of cells with fibronectin receptors such as mesenchymal stem cells (MSCs) and ECs.

Laminin: Essential for the growth and differentiation of neural stem cells and myogenic cells. Frequently used in neural and muscular differentiation cultures.

Matrigel: A gelatinous protein mixture rich in ECM components, Matrigel promotes cell proliferation and differentiation and is commonly used in three-dimensional (3D) cultures of tumor and stem cells.

Gelatin: A general-purpose adhesion enhancer that is particularly suitable for fibroblasts, ECs, and smooth muscle cells.

iMatrix: A high-performance recombinant fragment of laminin used as a coating material to support cell adhesion, proliferation, and lineage-specific differentiation. It is widely applied for the xeno-free culture of human-induced pluripotent stem cells (iPSCs) and MSCs for maintaining pluripotency or inducing differentiation. Compared with Matrigel, iMatrix offers greater consistency and clinical compatibility, making it suitable for Good Manufacturing Practices (GMP)-compliant cell manufacturing. It is also used for the induction of neural and endothelial lineages, where it enhances axon extension and expression of lineage-specific markers.

Cadherin-based coatings (e.g., E-cadherin-Fc): Cadherin is a calcium-dependent transmembrane glycoprotein involved in cell–cell adhesion. Coating culture dishes with cadherins can preserve the undifferentiated state of embryonic stem cells and iPSCs or support differentiation into specific lineages such as neurons and hepatocytes. A notable example is the use of the E-cadherin extracellular domain fused with the Fc region of IgG (E-cadherin-Fc). The Fc domain binds firmly to the polystyrene surfaces of culture plates, while the E-cadherin portion facilitates homophilic binding with cell surface cadherins. This system enables gentle, enzyme-free detachment of cells by chelating calcium with EDTA, making it especially useful for sensitive stem and progenitor cells. This technique, developed by Akaike and colleagues, is referred to as "cadherin matrix engineering" and represents an innovative biomaterials approach for cell recognition and tissue engineering [16, 17].

3.2.2 Selecting Culture Media for Stable Human Cell Lines

Establishing stable human cell lines requires not only proper tissue sourcing and handling but also careful optimization of the culture medium to support long-term cell proliferation, phenotypic stability, and resistance to stress-induced senescence. While conventional basal media are usually acceptable and widely used, such as Dulbecco's Modified Eagle Medium (DMEM) and DMEM/F12 (DMEM supplemented with the protein-free nutrient mixture F12 originally developed for serum-free culture of Chinese Hamster Ovary cells), successful cell line establishment and differentiation are highly sensitive to the presence and concentrations of various supplements such as serum, free amino acids, and buffering agents.

Historically, EC and MSC lines have been established in serum-supplemented basal media, commonly DMEM or DMEM/F12 with 10%–20% fetal bovine serum (FBS). Standard additives such as nonessential amino acids (NEAAs), GlutaMAX (a stabilized form of L-glutamine), and antibiotics (e.g., penicillin/streptomycin) are routinely included to enhance cell viability, minimize batch variability, and prevent contamination. For lineage-specific differentiation, particularly toward neural or epithelial lineages, additional factors such as insulin, transferrin, and selenium, retinoic acid, and various steroid hormones are incorporated.

3.2.2.1 Evidence from Our Laboratory Studies

Tamaki et al. [1] demonstrated that human dental pulp stem cells (DPSCs), periodontal ligament stem cells (PDLSCs), apical papilla stem cells, dental follicle stem cells, and bone marrow-derived mesenchymal stem cells could be successfully expanded in DMEM/F12 supplemented with 15% FBS, GlutaMAX, and antibiotics under normoxic conditions (20% O_2) for at least three passages without karyotypic instability or loss of multipotency [1].

Ishikawa et al. [13] reported the establishment of a benign meningioma cell line (TKB-MEN2) under low oxygen (3% O_2) using DMEM/F12 containing 10% FBS, NEAAs, GlutaMAX, and penicillin/streptomycin. This formulation supported natural immortalization and long-term propagation (>20 passages) without the need for introducing exogenous oncogenes, likely due to reduced oxidative stress and improved metabolic homeostasis under low-oxygen tension [13].

Takaoka et al. [10, 12] developed a protocol for neural induction from DPSCs using a two-step culture system. Cells were initially expanded in DMEM/F12 with 10%–15% FBS and then exposed to a neural induction medium (NIM) containing retinoic acid, nerve growth factor (NGF), dexamethasone, ascorbic acid, insulin, estradiol, progesterone, and thyroxine (T4). Cells were then maintained on Matrigel-coated dishes using Neurobasal medium supplemented with B27, epidermal growth factor (EGF), and fibroblast growth factor (FGF), facilitating the differentiation of DPSCs into neural precursors and mature neurons [10, 12].

In a related protocol, Matsumura et al. [8] generated dental pulp-derived induced neural cells (DP-iNCs) from DPSCs without genetic modification. After initial expansion in DMEM/F12 supplemented with 10% FBS, GlutaMAX, NEAAs, and antibiotics, neural induction was initiated by incubation in NIM supplemented with retinoic acid, steroid hormones, NGF, insulin, and ascorbic acid. Weakly adherent spheroid-forming cells were isolated and replated on Matrigel-coated culture plates, followed by maintenance culture in Neurobasal Plus medium supplemented with B27 (a mixture of antioxidant enzymes, vitamins, and fatty acids designed to support neuronal survival and the proliferation of neuronal cell lines), EGF, and FGF. These cells exhibited functional neural properties, including calcium signaling and electrophysiological responsiveness [8].

Media formulation is pivotal not only for initial survival and expansion but also for preserving cellular phenotype preventing senescence and ensuring lineage fidelity during differentiation. Tailoring the oxygen levels (such as lower levels for tumor-derived cells or MSCs) and adjusting the serum content can further enhance the culture success and cellular stability. Based on our experience establishing and maintaining both tumor-derived and stem cell-derived lines, we propose a stepwise strategy for media optimization.

1. Begin with a robust basal medium (e.g., DMEM/F12 + 10%–15% FBS + NEAAs + GlutaMAX) known to support the proliferation of a wide range of cell types for initial expansion.
2. Transition to a lineage-specific or differentiation medium supplemented with native hormones, growth factors, and matrix components.
3. Adjust microenvironmental parameters, including oxygen and ECM component composition, for stabilization and maturation.

This rational, adaptable approach facilitates the establishment of stable and functionally competent human cell lines suitable for both in vitro studies and translational research.

3.2.3 Optimization and Consistency of Culture Conditions

Maintaining consistent and optimized culture conditions is essential for preserving the phenotypic and genomic integrity of cultured cells. Variations in temperature, CO_2 concentration, pH, and medium composition can impose cellular stress, potentially leading to genomic instability, oxidative damage, and undesired phenotypic shifts. Therefore, long-term culture success depends not only on precise media formulation but also on the stability and reproducibility of the complete culture environment. Indeed, the shift toward serum-free culture was driven by documented phenotypic variations observed when using serum from different manufacturers and even between different lots from the same manufacturer.

3.2.3.1 Oxygen Concentration as a Critical Parameter

Among the environmental variables, oxygen concentration has a major influence on cell behavior and viability. Although standard cell culture conditions typically involve atmospheric oxygen levels (~20% O_2), this does not reflect the physiological microenvironment found within human and animal tissues, where oxygen tensions range from 1% to 5% depending on the organ and cell type. Moreover, exposure to atmospheric levels of oxygen can be detrimental to certain cell types by inducing oxidative stress, DNA damage, and premature senescence. For this reason, physiologically relevant hypoxic conditions (1–5% O_2) are frequently adopted for the culture of stem cells, neural cells, and embryonic cells and have been shown to enhance cell survival, maintain multipotency, and modulate differentiation capacity. For example, in our previous study on benign meningioma (TKB-MEN2) cells, long-term culture under 3% O_2 promoted natural immortalization and reduced cellular senescence without the need for transgenic manipulation [13]. In contrast, these same cells cultured under atmospheric oxygen levels ("normoxia") showed morphological deterioration, elevated secretion of the proinflammatory cytokine interleukin (IL)-6, and increased activity of the senescence marker SA-β-galactosidase. Similarly, studies on MSCs and neural precursor cells have reported enhanced survival and more reliable lineage-specific gene expression profiles under low-oxygen conditions that more closely mimic the in vivo stem cell niche. In cancer research, low-oxygen tension is also employed to simulate the hypoxic tumor microenvironment, as some tumor-derived cell lines exhibit abnormal proliferation and stress resistance under such conditions.

Optimizing the oxygen tension based on the cell type and research objective is a fundamental aspect of advanced cell culture. For the routine propagation of immortalized lines, 20% O_2 may be adequate, but for sensitive cell types such as stem cells, neural cells, or primary isolates, hypoxic culture may be superior for maintaining cell function, identity, and experimental relevance. The implementation of controlled oxygen environments should thus be considered an essential parameter in modern cell culture strategies [18].

3.2.4 Case Study: Establishment of a Benign Meningioma Cell Line Under Hypoxic Conditions

As a practical example of stable cell line development, we highlight our recent establishment of a spontaneously immortalized cell line derived from a WHO grade I benign meningioma [13].

Meningiomas are the most common primary brain tumor, accounting for approximately 41% of central nervous system tumors in the United States. Approximately 80% of these are classified as WHO grade I and are typically considered benign and curable via surgical resection. However, in cases where the tumor arises at the skull base or in proximity to critical vascular structures, complete excision is often difficult. For these cases, radiation therapy may be an effective alternative, but recurrence may occur. These clinical challenges underscore the need for novel therapeutic strategies that consider recurrence risk even for benign meningiomas. Cell lines stably maintaining core clinically relevant phenotypes are essential for foundational studies. Despite the prevalence of grade I meningiomas, however, there are few stable cell lines derived from these tumors, and most in vitro models rely on the introduction of immortalizing genes such as hTERT, SV40 large T antigen, or HPV E6/E7 to bypass replicative senescence. Therefore, the establishment of benign meningioma cell lines without exogenous genetic manipulation remains critical for developing physiologically relevant models.

In our study, we established a novel cell line, designated TKB-MEN2, from a transitional-type WHO grade I meningioma resected from a 71-year-old male patient with no evidence of metastasis. The cells were cultured under hypoxic conditions (3% O_2) without transfection of immortalizing genes (Fig. 3.1). Remarkably, the TKB-MEN2 line demonstrated stable proliferation over 20 passages without exhibiting signs of senescence or acquiring genetic hallmarks of malignancy such as TERT promoter mutations or CDKN2A/2B deletions.

In contrast, when cultured under normoxic (20% O_2) conditions, these cells exhibited evident signs of senescence, including morphological alterations, increased IL-6 secretion, and elevated senescence-associated β-galactosidase (SA-β-gal) activity. These findings emphasize the detrimental impact of relative hyperoxia (e.g., atmospheric oxygen levels) on the stability of benign tumor-derived cells and highlight hypoxia as a protective and stabilizing factor for cell line establishment. This study describes one of the first benign meningioma cell lines demonstrating spontaneous immortalization without genetic modification and underscores the value of physiologically relevant oxygen conditions for supporting long-term cell viability, suppressing senescence, and preserving genetic stability in culture [13].

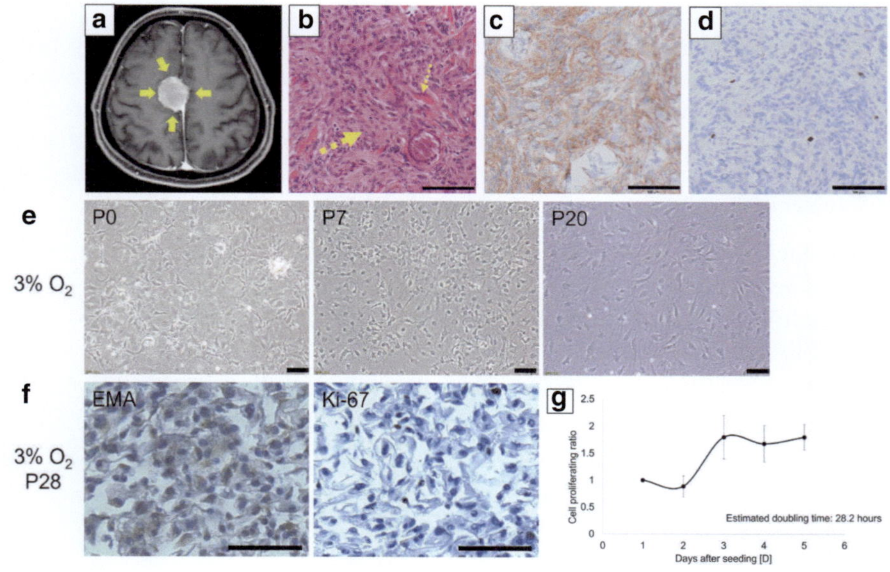

Fig. 3.1 Clinical and histological features of the source meningioma used to derive the TKB-MEN2 cell line. (**a**) T1-weighted gadolinium-enhanced MRI showing a well-demarcated, intensely enhancing tumor attached to the falx cerebri. (**b**) Hematoxylin and eosin staining revealing a syncytial growth pattern (bold dotted arrows) and fibrous tissue structure (thin dotted arrows). (**c**) Immunohistochemical staining for the epithelial membrane antigen (EMA). (**d**) Approximately 1% of the tumor cells were positive for the proliferation marker Ki-67. Scale bars: 100 μm. (**e**) Phase-contrast images of TKB-MEN2 cells cultured under 3% O_2 (hypoxic conditions) at passage 0 (P0), 7 (P7), and 20 (P20), confirming maintained proliferation. (**f**) Immunocytochemistry at passage 28 (P28) confirmed EMA positivity and low Ki-67 expression in TKB-MEN2. Scale bars: 100 μm. (**g**) Proliferation curve of TKB-MEN2 cells, yielding an estimated doubling time of 28.2 h between 48 and 72 h after seeding. (*Reproduced from: Ishikawa T, et al. Establishment of a novel benign meningioma cell line spontaneously immortalized under hypoxic conditions. Hum Cell. 2024; 38:22. © Springer Nature. Reproduced with permission*)

3.3 Passaging Strategies for Cultured and Established Cell Lines

Proper subculture (passaging) of established cell lines and primary cells is essential for maintaining functional integrity and genomic stability. Without careful management, cellular characteristics can drift, and genetic abnormalities may accumulate during prolonged in vitro expansion. This section outlines the key considerations for passaging, including optimal timing, enzymatic dissociation techniques, and strategies for selecting desirable subpopulations. By implementing the following approaches, researchers can enhance cell stability and preserve the intended biological functions over long-term culture.

3.3.1 Maintaining Low-Passage Numbers

As the passage number increases, cultured cells are more likely to acquire genetic alterations and phenotypic changes. This "cellular fatigue" phenomenon is particularly pronounced in naturally immortalized or primary-derived lines, where prolonged in vitro maintenance may lead to karyotypic instability, loss of differentiation potential, or aberrant proliferation. To reduce the risk of progressive instability, we recommend the following.

- Use low-passage cells for experimental work, ideally below passage 10–15 after cell line establishment, depending on the cell type.
- Establish early-stage cell banks by cryopreserving aliquots of cells shortly after establishment or isolation.
- Return to early-passage stocks when phenotypic drift is suspected or after extended use.

Maintaining low-passage numbers improves reproducibility and helps prevent cumulative damage that may compromise the validity of downstream assays [1, 8, 14, 15, 19].

3.3.2 Selecting Subpopulations with Low Proliferative Activity

Routine passaging of confluent cultures often favors highly proliferative subclones, which can lead to hyperploidy, functional heterogeneity, and loss of the original cell traits. To avoid these changes, it is essential to subculture less proliferative, morphologically stable cells. One effective method for retaining the predominance of subclones with the original characteristics is low-density seeding (sparse plating) as follows:

- Seed cells at low density (e.g., 1×10^4 to 5×10^3 cells per 10-cm dish).
- Allow 2–3 weeks in culture for individual colony formation.
- Select colonies composed of cells with dark cytoplasm under phase-contrast microscopy as this appearance is typically indicative of cytoplasm rich in organelles. These cells are often metabolically active, secretory, and maintain lineage characteristics.

This subculture selection method is based on cellular morphology, drawing parallels to epithelial differentiation models such as for eccrine sweat glands. Clear cells in sweat glands are rich in glycogen but poor in organelles, resulting in an electron-lucent cytoplasm, whereas darker cells contain abundant secretory granules and exhibit active protein secretion. By analogy, dark-appearing colonies in cell culture may reflect a more functional, stable phenotype suitable for long-term use [20, 21].

3.3.3 Enzymatic Dissociation for Cell Passaging

The choice of enzymatic dissociation method is a critical determinant of successful cell passaging, as it directly influences cell viability, phenotypic retention, and surface marker preservation. Enzymes must be selected based on the cell type, adhesion characteristics, and sensitivity to proteolytic digestion. Commonly used enzymes include trypsin, collagenase, and dispase, each with distinct properties suited for specific applications. In some cases, it may be necessary to add a divalent cation chelator such as EGTA, while in others, EGTA alone may be sufficient. A well-calibrated dissociation strategy minimizes cellular damage and ensures efficient recovery.

Trypsin: Trypsin is the most widely used enzyme for the routine subculture of adherent cells. It cleaves peptide bonds in surface proteins, thereby facilitating detachment from the substrate. Trypsin is broadly applied to fibroblasts, epithelial cells, and endothelial cells. However, its nonspecific proteolytic activity can damage sensitive cell types or surface antigens. Therefore, its use should be limited to 1–5 min at 37 °C and promptly terminated with serum-containing medium or a dedicated trypsin inhibitor [14].

EDTA (Ethylenediaminetetraacetic Acid): EDTA is a chelating agent that sequesters divalent cations (Ca^{2+}, Mg^{2+}), thereby disrupting calcium-dependent cell adhesion molecules. It is commonly used in combination with trypsin (Trypsin-EDTA solution) to reduce proteolytic stress. In cadherin-coated culture systems, EDTA alone can effectively dissociate cells via the disruption of homophilic cadherin interactions. EDTA is generally gentler than trypsin but must still be used at optimized concentrations and durations to avoid adverse effects on cellular integrity [19].

Collagenase: Collagenase specifically degrades collagen, a major component of the extracellular matrix (ECM). It is particularly useful for primary cell isolation from solid tissues such as skeletal muscle, cartilage, or adipose tissue, where dense ECM limits the efficacy of trypsin. Collagenase acts more selectively and is less damaging during extended incubation, making it suitable for ECM-rich tissues or trypsin-sensitive cells. However, use in combination with other enzymes requires careful monitoring of potential synergistic or antagonistic effects.

Dispase: Dispase targets intercellular adhesion molecules without extensively cleaving surface proteins. Therefore, dispase enables gentle detachment of cells and is suitable for epithelial and stem cell cultures where surface marker preservation is essential. Dispase is also less aggressive than trypsin and thus ideal for applications requiring intact membrane proteins, although its detachment efficiency may be comparatively lower [14, 19].

3.4 Cell Banking, Authentication, and Routine Monitoring

Even with optimized protocols for culture, passaging, and differentiation, ensuring the long-term phenotypic stability of experimental cells and the reliability of experimental results requires the implementation of regular quality control measures. Establishing validated cell stocks and performing routine authentication and molecular monitoring are essential for experimental reproducibility and scientific integrity.

3.4.1 Regular Cell Authentication and Identity Profiling

Periodic authentication is critical for identifying genetic drift, spontaneous mutations, and particularly cross-contamination, which remains a common issue in shared or long-term laboratory cultures. A widely accepted method for confirming cellular identity is short tandem repeat profiling, which allows for comparison against reference databases to verify the origin and uniqueness of a given cell line.

Frequent authentication helps to

- Detect misidentified or contaminated cell lines.
- Confirm the integrity of in-house derived or modified cell lines.
- Ensure consistency across experimental replicates and laboratories.

The inclusion of authentication procedures in standard laboratory workflows is no longer optional but a necessity for credible cell-based research [19].

3.4.2 Molecular-Level Monitoring of Genetic and Phenotypic Stability

In addition to identity profiling, the molecular characterization of cell populations provides a means to detect early signs of genomic instability, epigenetic changes, and phenotypic drift. The recommended approaches include the following:

- Periodic gene expression analysis (e.g., RT-qPCR or RNA-seq) for lineage markers.
- Protein profiling using flow cytometry or immunocytochemistry.
- Screening for mutations or copy number variations in key genes related to proliferation, apoptosis, and differentiation.

Routine molecular monitoring allows for early intervention before the experimental outcomes are compromised. For example, the detection of downregulated neural markers or upregulated senescence-associated genes could be used as a sign to revive lower-passage stocks or optimize the culture conditions. Maintaining this level of surveillance minimizes the risk of unanticipated variability, enhances reproducibility, and contributes to the long-term success of stem cell and tumor-derived cell models [19, 22].

3.5 Representative Procedures for the Establishment of Stable Human Cell Lines

This section summarizes protocols and experimental results from our laboratory to illustrate how the judicious selection of culture conditions, passaging strategy, enzymatic dissociation, and authentication can enhance phenotypic stability and improve study replicability. These examples include the establishment of stable cell lines derived from benign brain tumors and oral mesenchymal tissues, the directed differentiation of these cell lines into neural lineages, and applications in regenerative therapies for ischemic stroke and neural injury.

We have developed and characterized several cell lines under the following defined and physiologically relevant conditions.

3.5.1 Human-Derived Tsukuba Chordoma 1 Cells

Chordoma is a rare, low-grade malignant tumor arising from remnants of the embryonic notochord. It is resistant to conventional therapies and has garnered attention as a potential target for differentiation-inducing treatments. However, the number of established human chordoma cell lines remains extremely limited, and there is a particular scarcity of stable lines that retain the pathophysiological properties of primary tumor cells. Kino et al. [9] established the Tsukuba Chordoma 1 (TSK-CHO1) line from a clinical specimen and comprehensively evaluated the biological effects of its conditioned medium on other cell types [9].

Primary chordoma cells were cultured in DMEM/F12 supplemented with 10% FBS under standard conditions (37 °C, 5% CO_2) and passaged using 0.25% trypsin/EDTA. Highly adherent tumor cells were selectively expanded, resulting in the establishment of a stable and continuously proliferating cell line. These TSK-CHO1 cells exhibited a polygonal to spindle-shaped morphology with moderately enlarged nuclei and abundant cytoplasm, expressed key chordoma-associated transcription factors such as Brachyury (T) and SOX9, and maintained proliferative activity through multiple passages (Fig. 3.2). Analysis of the conditioned medium revealed the presence of secreted proteins, including Brachyury, SOX9, and cartilage oligomeric matric protein, suggesting the potential to induce the chondrogenic differentiation of recipient cells.

Given this clinical chordoma origin and documented preservation of physiologically relevant features, the TSK-CHO1 line represents a valuable resource for chordoma research. The capacity to secrete chondrogenic differentiation factors also supports its utility as a model to study differentiation therapy for cancer and cartilage regeneration [9].

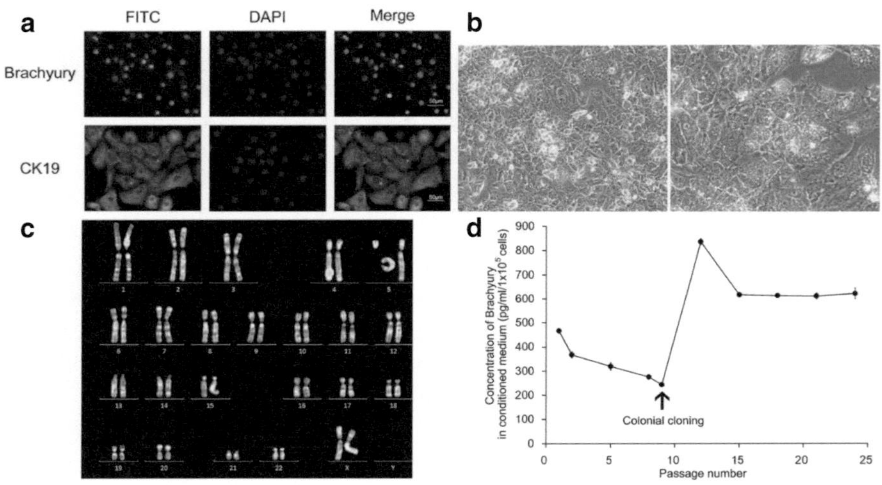

Fig. 3.2 Characterization of TSK-CHO1 cells. (**a**) Immunofluorescence staining of TSK-CHO1 cells. Strong nuclear expression of Brachyury and robust cytoplasmic expression of CK19 were observed. FITC, fluorescein isothiocyanate. Scale bar = 50 μm. (**b**) Phase-contrast microscopy of TSK-CHO1 cells showing the characteristic physaliferous morphology. Scale bars: 200 μm (left) and 50 μm (right). (**c**) Representative karyogram of a metaphase spread from TSK-CHO1 cells showing multiple chromosomal abnormalities identified by G-banding analysis. (**d**) Concentration of Brachyury protein in the conditioned medium at various passage numbers. A marked increase was observed following clonal selection at passage 10 (arrow). (*Reproduced from: Kino H, et al. Inducing substances for chondrogenic differentiation of dental pulp stem cells in the conditioned medium of a novel chordoma cell line. Hum Cell. 2022; 35:745–55. © Springer Nature. Reproduced with permission*)

3.5.2 Neural Cells Derived from Human DPSCs [12]

Takaoka et al. [12] developed a neural induction protocol for human DPSCs and analyzed the molecular characteristics of the induced neural cell populations [10, 12]. Human DPSCs were isolated from the dental pulp tissue of extracted human teeth, minced, and grown in adherent culture under static conditions. Migrated cell colonies were selected and expanded in DMEM/F12 supplemented with 10% FBS, GlutaMAX, NEAAs, and antibiotics at 37 °C under a 4.7% CO_2 atmosphere. Upon reaching confluence, the cells were passaged at a 1:3 ratio.

For neural induction, 1×10^4 DPSCs were reseeded in 60-mm dishes and cultured for approximately 2 weeks in a medium composed of DMEM/F12 supplemented with 5% FBS, glutamic acid, retinoic acid, ascorbic acid, insulin, dexamethasone, estradiol, progesterone, NGF, and T4 (induction medium). Under these induction conditions, colonies exhibiting a neural-like morphology emerged, and the expression of neural progenitor markers was confirmed. Gene expression profiling using microarray analysis and single-cell RNA sequencing (scRNA-seq) revealed upregulation of neural development-related genes such as FABP7 and ZIC1, as well as activation of axon guidance pathways. Cell clustering based on scRNA-seq identified populations corresponding to neural progenitors, neurons, and astrocytes, as well as a minor fraction of nonneural cell types. Pseudotime trajectory analysis further revealed a developmental pathway wherein neural progenitors first differentiated into neurons, followed by a transition into astrocytes, successfully mimicking in vivo neurogenesis (Fig. 3.3).

This neural induction strategy offers a highly reproducible and efficient method for generating neural progenitor and mature neural cells from DPSCs. Compared to iPSC-based approaches, it presents fewer ethical and safety concerns, making it a promising alternative source of neural cells for research and therapeutic applications targeting neurological diseases [10, 12].

3.5.3 Establishment of Dental Pulp-Induced Neural Cells

Matsumura et al. [8] reported the successful establishment of Dental Pulp-induced Neural Cells (DP-iNCs) from human DPSCs without the use of genetic modification [8]. Primary cultures were initiated in DMEM/F12 supplemented with 10% FBS, GlutaMAX, NEAAs, antibiotics, and fungizone and maintained at 37 °C under a 5% CO_2 atmosphere. Once the cells reached near confluence, the medium was switched to NIM containing retinoic acid, ascorbic acid, insulin, steroid hormones (progesterone, estradiol, dexamethasone), NGF, and T4. Cells were maintained under these conditions for approximately 2 weeks. During the induction process, weakly adherent spheroid-forming cells emerged and were selectively harvested via pipetting. These spheroids were enzymatically dissociated and replated on Matrigel-coated dishes for further expansion under the same induction

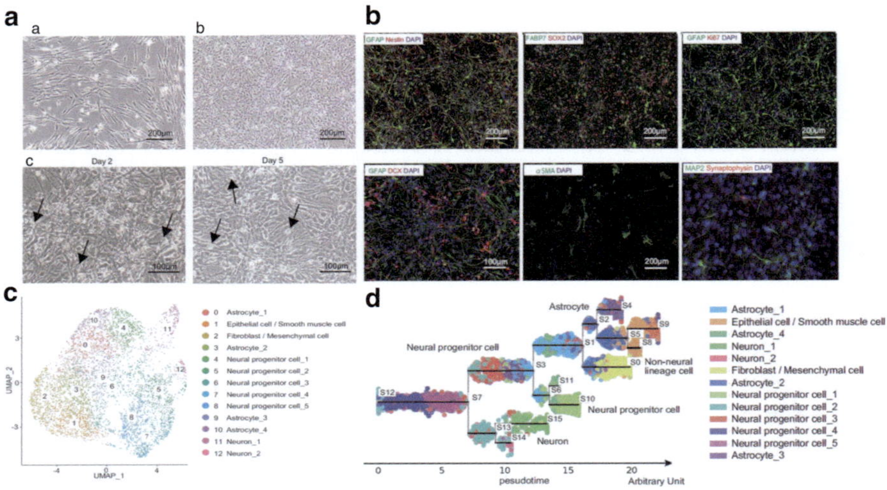

Fig. 3.3 Morphological changes during neural induction culture of DPSCs. (**a**) a, Phase-contrast image of undifferentiated DPSCs. b, Ni-DPSCs on day 2 of induction (2 days in vitro, DIV) showing early morphological transformation. c, High-magnification phase-contrast images of Ni-DPSCs at 2 and 5 DIV. d, Numerous small, round cells appeared in the upper layer (black arrows), likely neural progenitor-like populations emerging during early neural commitment. (**b**) Immunocytochemical characterization of Ni-DPSCs. a, At 2 DIV, Ni-DPSCs expressed early neural and stem cell markers, including nestin, glial fibrillary acidic protein (GFAP), SOX2, FABP7, Ki-67, and DCX, along with aSMA. b, At 5 and 10 DIV, expression of the mature neuronal markers MAP2 and synaptophysin was observed, indicating progressive neural differentiation. (**c**) Single-cell RNA sequencing analysis. Two-dimensional UMAP visualization revealed 13 distinct transcriptional clusters among Ni-DPSCs, indicating cellular heterogeneity and lineage specification. (**d**) Pseudotime trajectory analysis. Flat tree visualization generated by STREAM depicts differentiation pathways of neural progenitor clusters toward neuronal and astrocytic lineages. (*Reproduced from: Takaoka S, et al. Sequencing-based study of neural induction of human dental pulp stem cells. Hum Cell. 2024;37:1638–1648. © Springer Nature. Reproduced with permission*)

conditions. Expanded cells were then maintained in Neurobasal Plus medium supplemented with B27, EGF, and FGF (both at 20 ng/mL). Cells from passages 4–6 were used primarily for the experimental analyses.

The resulting DP-iNCs exhibited a CD184(−)/CD44(−)/CD15^low/CD24(+) immunophenotype, expressed neural markers, and demonstrated functional properties such as calcium signaling and electrical activity as assessed by fluorometric imaging and multielectrode array recordings, respectively. This protocol offers a robust method for generating functional neural-like cells from DPSCs without introducing exogenous genes and has also demonstrated therapeutic efficacy in a mouse model of cerebral ischemia [8] (Fig. 3.4).

Funding This work was supported by JSPS KAKENHI (Grant Number: JP22K09278, JP23K18337, JP23K24440, JP23K27707, JP24K02527) to AM and HI, JST FOREST Program (Grant Number: JPMJFR2112) to AM.

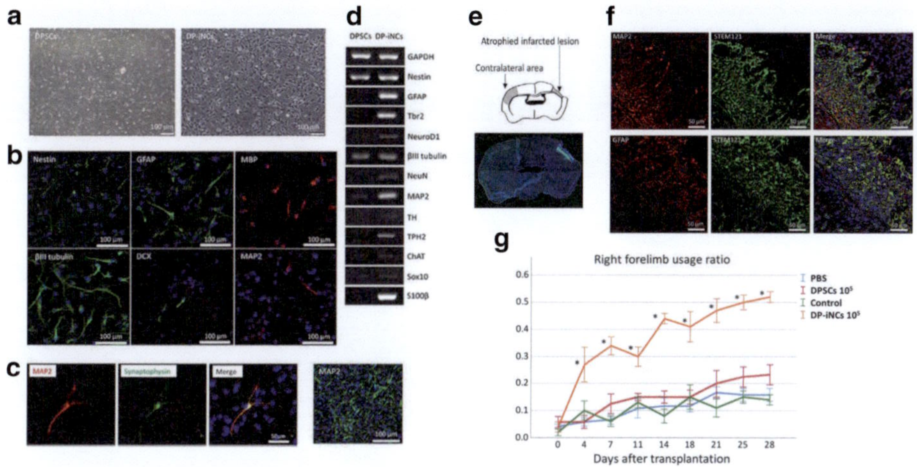

Fig. 3.4 Characteristics and functional potential of dental pulp-derived induced neural cells (DP-iNCs). (**a**) Morphological comparison between DPSCs and DP-iNCs. Phase-contrast microscopy images of DPSCs revealed a spindle-shaped morphology, while DP-iNCs displayed multipolar, axon-like projections characteristic of neural lineage cells. (**b**) Immunocytochemical analysis of DP-iNCs 7 days after subculture. Positive staining was observed for neural and glial markers including nestin, glial fibrillary acidic protein (GFAP), myelin basic protein (MBP), doublecortin (DCX), βIII-tubulin, and microtubule-associated protein 2 (MAP2), indicating progression toward neural differentiation. (**c**) Immunofluorescence triple-labeling for MAP2 (red), synaptophysin (green), and DAPI (blue). The colocalization of MAP2 and synaptophysin demonstrated the presence of mature neurons with presynaptic structures. At 1 month after subculture, DP-iNCs exhibited increased expression of these markers compared with the 7-day cultures shown in panel B. (**d**) Comparative mRNA expression profiles. Gene expression analysis revealed transcriptional differences between undifferentiated DPSCs and differentiated DP-iNCs, supporting neural lineage commitment. (**e**) In vivo distribution of transplanted DP-iNCs. At 4 weeks posttransplantation, DP-iNCs were found within deeper areas of infarcted brain lesions in ischemic stroke model mice. STEM121 (green) indicates human cytoplasmic protein expression, and DAPI (blue) marks the nuclei. (**f**) Immunohistochemistry of the transplanted cells. Colocalization of the human-specific marker STEM121 (green) with MAP2 (red, mature neurons) and GFAP (red, astrocytes), suggested multilineage integration of transplanted DP-iNCs into the host neural tissue. (**g**) Behavioral assessment of functional recovery. Mice transplanted with DP-iNCs showed significantly improved right forelimb usage compared with the PBS, DPSC-only, and control groups. Data are presented as mean ± SEM. *$P < 0.05$ by linear mixed model. (*Reproduced from: Matsumura H, et al., "Induced Neural Cells from Human Dental Pulp Ameliorate Functional Recovery in a Murine Model of Cerebral Infarction", Stem Cell Reviews and Reports, 2022; 18:595–608. © Springer Nature. Reproduced with permission*)

Conflicts of Interest Aiki Marushima is the inventor of the redox nanoparticles in reference [Hirata, Human Cell, 2023] and the representative director and shareholder of CrestecBio Inc. Hiroshi Ishikawa has no conflicts of interest to declare.

Ethical Approval and Informed Consent The articles of Matsumura, Stem Cell Rev Rep, 2022, Kino, Human Cell, 2022, Takaoka, Human Cell, 2024, and Ishikawa T, Human Cell, 2024 described in this chapter were approved by the University of Tsukuba Clinical Research Ethics Review Committee. Details of ethical approval and informed consent can be found in the original articles.

References

1. Tamaki Y, Nakahara T, Ishikawa H, Sato S. In vitro analysis of mesenchymal stem cells derived from human teeth and bone marrow. Odontology. 2013;101:121–32.
2. Takahashi H, Ishikawa H, Mataga I, Tanaka A. Establishment and characterization of human lingual squamous cell carcinoma cell lines designated Nialym derived from metastatic foci of lymph node, and Nialymx derived from transplanted tumor of Nialym cells. Hum Cell. 2015;28:143–53.
3. Nakahara T, Tominaga N, Toyomura J, Tachibana T, Ide Y, Ishikawa H. Isolation and characterization of embryonic ameloblast lineage cells derived from tooth buds of fetal miniature swine. In Vitro Cell Dev Biol Anim. 2016;52:445–53.
4. Ohyama A, Toyomura J, Tachibana T, Isonishi S, Takahashi H, Ishikawa H. Establishment and characterization of a clear cell carcinoma cell line, designated NOCC, derived from human ovary. Hum Cell. 2016;29:188–96.
5. Kawakami M, Ishikawa H, Tanaka A, Mataga I. Induction and differentiation of adipose-derived stem cells from human buccal fat pads into salivary gland cells. Hum Cell. 2016;29:101–10.
6. Tansriratanawong K, Ishikawa H, Toyomura J, Sato S. Establishment and characterization of novel epithelial-like cell lines derived from human periodontal ligament tissue in vitro. Hum Cell. 2017;30:237–48.
7. Takahashi H, Ishikawa H, Tanaka A. Regenerative medicine for Parkinson's disease using differentiated nerve cells derived from human buccal fat pad stem cells. Hum Cell. 2017;30:60–71.
8. Matsumura H, Marushima A, Ishikawa H, Toyomura J, Ohyama A, Watanabe M, Takaoka S, Bukawa H, Matsumura A, Matsumaru Y, Ishikawa E. Induced neural cells from human dental pulp ameliorate functional recovery in a murine model of cerebral infarction. Stem Cell Rev Rep. 2022;18:595–608.
9. Kino H, Akutsu H, Ishikawa H, Takano S, Takaoka S, Toyomura J, Hara T, Ishikawa E, Matsumaru Y, Bukawa H, Matsumura A. Inducing substances for chondrogenic differentiation of dental pulp stem cells in the conditioned medium of a novel chordoma cell line. Hum Cell. 2022;35:745–55.
10. Takaoka S, Uchida F, Ishikawa H, Toyomura J, Ohyama A, Watanabe M, Matsumura H, Marushima A, Iizumi S, Fukuzawa S, Ishibashi-Kanno N, Yamagata K, Yanagawa T, Matsumaru Y, Bukawa H. Transplanted neural lineage cells derived from dental pulp stem cells promote peripheral nerve regeneration. Hum Cell. 2022;35:462–71.
11. Hirata K, Marushima A, Nagasaki Y, Ishikawa H, Matsumura H, Mujagić A, Hirayama A, Toyomura J, Ohyama A, Takaoka S, Bukawa H, Matsumura A, Ishikawa E, Matsumaru Y. Efficacy of redox nanoparticles for improving survival of transplanted cells in a mouse model of ischemic stroke. Hum Cell. 2023;36:1703–15.
12. Takaoka S, Uchida F, Ishikawa H, Toyomura J, Ohyama A, Matsumura H, Hirata K, Fukuzawa S, Kanno NI, Marushima A, Yamagata K, Yanagawa T, Matsumaru Y, Ishikawa E, Bukawa

H. Sequencing-based study of neural induction of human dental pulp stem cells. Hum Cell. 2024;37:1638–48.
13. Ishikawa T, Matsuda M, Ishikawa H, Toyomura J, Ohyama A, Sakamoto N, Zaboronok A, Ishikawa E. Establishment of a novel benign meningioma cell line spontaneously immortalized under hypoxic conditions. Hum Cell. 2024;38:22.
14. Freshney RI. Culture of animal cells: a manual of basic technique and specialized applications. 7th ed. Oxford: Wiley-Blackwell; 2015.
15. Shalini M, Manisha S, Kumar A. Animal cell culture: principles and practice. Berlin: Springer; 2023.
16. Haque MA, Nagaoka M, Hexig B, Akaike T. Artificial extracellular matrix for embryonic stem cell cultures: a new frontier of nanobiomaterials. Sci Technol Adv Mater. 2010;11:014106.
17. Shibata-Seki T, Nagaoka M, Goto M, Kobatake E, Akaike T. Direct visualization of the extracellular binding structure of E-cadherins in liquid. Sci Rep. 2020;10:17044.
18. Packer L, Ong J, Halliwell B. Hypoxia in cell culture and its effects on cellular functions. Berlin: Springer; 2018.
19. Masters JR, Stacey GN. Changing medium and passaging cell lines. Nat Protoc. 2007;2:2276–84.
20. Ross MH, Pawlina W. Histology: a text and atlas: with correlated cell and molecular biology. 8th ed. Philadelphia, PA: Wolters Kluwer; 2020.
21. Sato K, Kang WH, Saga K, Sato KT. Biology of sweat glands and their disorders. I. Normal sweat gland function. J Am Acad Dermatol. 1989;20:537–63.
22. Ryan JA. Understanding and managing cell culture contamination. New York: Corning Incorporated; 1994.

Chapter 4
Recent Advances in Human Type II Alveolar Epithelial Cell Research

Takunori Ogawa

Abstract Refractory pulmonary diseases such as idiopathic pulmonary fibrosis and chronic obstructive pulmonary disease (COPD) currently affect a significant number of patients worldwide. These diseases not only negatively impact global health but also strain the global economy. In recent years, anti-fibrotic agents for pulmonary fibrosis and inhaled bronchodilators for COPD have emerged. However, their efficacy against these diseases remains limited. The development of novel effective therapies for these intractable pulmonary diseases requires an understanding of the pathogenesis of the disease, which is still not fully elucidated. This section describes the latest developments in pulmonary epithelial cells, with a particular focus on alveolar epithelial cells and their culture methods.

Keywords Alveolar epithelial cells · Culture media · Organoid

4.1 Introduction: Human Lung Structure

The human respiratory system is composed of two closely interrelated systems: the pulmonary epithelium, which carries air to the alveoli, and the vasculature, which carries blood through the alveoli for gas exchange. The pulmonary epithelium is composed of tissue stem cells and differentiated cells, with their composition varying based on the specific area covered by the pulmonary epithelial cells. These cells are broadly classified into airway (tracheal/bronchiolar) and alveolar types. Human airways are lined by a pseudostratified epithelium, where each cell is in contact with a basement membrane [1]. The airway epithelium primarily comprises basal, secretory (club, mucous, and serous subtypes), and ciliated cells [2]. Basal cells serve as

T. Ogawa (✉)
Division of Infectious Diseases and Respiratory Medicine, Department of Internal Medicine, National Defense Medical College, Tokorozawa, Saitama, Japan
e-mail: taku0107@ndmc.ac.jp

N. Shinomiya, Y. Shimada (eds.), *Applications of Cell Culture*, Current Human Cell Research and Applications, https://doi.org/10.1007/978-981-95-0587-6_4

stem cells within the airway epithelia that can self-renew post-injury and differentiate into other airway epithelial cells, including ciliated and secretory cells. The secretory products of secretory cells play a crucial role in maintaining airway homeostasis, and their dysregulation contributes to the development of many respiratory diseases. The cilia on ciliated cells are specialized organelles that beat in metachronal waves, propelling pathogens and inhaled particles trapped in the mucous layer out of the airways [3].

The alveolar epithelium consists of type I and type II alveolar epithelial cells (AT1 and AT2, respectively) [4]. AT1 cells are thin, flat epithelial cells that cover approximately 95% of the alveolar surface and are involved in gas exchange. AT2 cells are cuboidal epithelial cells that synthesize surfactants, store them in lamellar bodies, and secrete them into the alveolar lumen. Pulmonary surfactants covering the alveoli are essential to reduce alveolar surface tension and prevent alveolar collapse at the end of expiration [5]. Interestingly, AT2 cells also function as tissue stem cells within the alveolar region, possessing the ability to self-renew and differentiate into AT1 cells [6, 7].

4.2 AT2 Cells in Refractory Pulmonary Diseases

Pulmonary epithelial cells serve not only as the first line of physical defense against foreign substances but also play a crucial role in various respiratory immune responses [8, 9]. Among these cells, AT2 cells stand out as stem cells responsible for tissue repair of damaged alveolar epithelium. They are also implicated in the pathogenesis of distal lung diseases, including idiopathic pulmonary fibrosis (IPF) and chronic obstructive pulmonary disease (COPD).

IPF, a progressive, irreversible, and refractory pulmonary disease, carries a poor prognosis—with a median survival of 3.8 years in US adults aged 65 years and older [10]. Researchers have suggested that the loss of normal AT2 cell function and the acquisition of a pro-fibrotic phenotype may contribute to IPF development [11]. Notably, mutations in *SFTPC*, a characteristic gene of AT2 cells, are known to cause fibrosis of the lungs [12]. In patients with IPF, senescence and apoptosis are enhanced in AT2 cells. Their telomere length is shorter in fibrotic regions compared to that in non-fibrotic regions [13]. Cellular senescence markers were upregulated in distal lung epithelial cells in patients with IPF [14, 15]. Smoking and aging—well-established risk factors for IPF—contribute to telomere shortening [16]. Furthermore, AT2 cells in the IPF lungs exhibit DNA fragmentation, a finding suggestive of cell apoptosis [17]. Mitochondrial stress and endoplasmic reticulum (ER) stress in AT2 cells were also involved in IPF pathogenesis. AT2 cells in IPF lungs show accumulation of dysmorphic and dysfunctional mitochondria associated with elevated expression of ER stress markers [18]. Additionally, injured AT2 cells in both bleomycin-injured mouse lungs and human IPF lungs activate fibroblasts by secreting profibrotic mediators [19–22].

COPD is a common respiratory condition, with an estimated prevalence of approximately 10% of adults worldwide. However, the exact prevalence varies based on survey methods, diagnostic criteria, and statistical approaches [23, 24]. COPD stands as a leading cause of mortality worldwide, accounting for an estimated three million deaths annually (2015, [25]). COPD is a disease characterized by progressive airflow limitation, stemming from bronchitis, small airway remodeling, and emphysema. Cigarette smoking is a significant contributor to COPD pathogenesis. It disrupts normal surfactant production and secretion in AT2 cells [26, 27]. Notably, patients with COPD had significantly lower surfactant protein D (SP-D) levels compared to healthy smokers [28]. In mice, SP-D deficiency leads to spontaneous airspace enlargement [29, 30]. Based on these results, it is likely that smoking-induced abnormalities in surfactant secretion from AT2 cells are associated with the development of COPD. In the lungs of patients with COPD, an abnormal inflammatory response persists long after smoking cessation [31, 32]. In an in vivo animal model, smoking was found to activate the NF-κB pathway in AT2 cells [33]. Furthermore, inflammatory AT2 cell populations, expressing chemokines such as CXCL1 and CXCL8, are increased in the lungs of patients with COPD [34].

Altogether, the study of AT2 cells is necessary to elucidate the pathogenesis of refractory pulmonary diseases.

4.3 Issues Arising from Different Methods of Studying Refractory Pulmonary Diseases

4.3.1 Animal Models for Pulmonary Disease Research

Animal models are widely used to study the pathogenesis of pulmonary diseases. They provide valuable insights into the complex immune networks and regenerative processes occurring in vivo, which are difficult to replicate in simple in vitro experimental systems. Additionally, testing experimental results in animal models before translating them to humans allows for partial confirmation of efficacy and safety. However, most animal models do not fully reflect the pathobiology associated with tissue damage, cellular inflammation, immune response, tissue repair, and fibrosis observed in the lungs of patients with intractable pulmonary diseases [35]. Bleomycin can cause pulmonary fibrosis in a short period and is, therefore, most often used in animal models of pulmonary fibrosis [36]. However, the pathology is not fully consistent with usual interstitial pneumonia, which characterizes IPF lung. Furthermore, these animal models often develop fibrosis resulting from acute inflammation that eventually resolves, in contrast to the chronic, progressive disease observed in patients with IPF. Animal models overexpressing cytokines such as TGF-β, TGF-α, and IL-13 are also used to study pulmonary fibrosis [36]; however, they are primarily useful for examining the downstream signals of these cytokines in the pathogenesis of pulmonary fibrosis rather than the upstream ones.

Turning to COPD, its pathogenesis is complex, and no single animal model completely replicates its multifaceted pathophysiology [37]. As an example, intratracheal administration of elastolytic enzymes (such as the porcine pancreatic elastase, human neutrophilic elastase, and papain) causes inflammatory cell infiltration and subsequent emphysematous changes in the lungs. The advantage of the elastase-induced emphysema model lies in its rapid disease onset and ease of measuring functional changes, making it one of the most commonly used COPD models. The disadvantage of the elastase model is that the functions of elastase and cigarette smoke in COPD emphysema may be mediated by very different pathophysiological mechanisms [38, 39].

Animal models using cigarette smoke are also used. Although they mimic the repeated inhalation of smoke and the onset of COPD in humans, they require 4–6 months for the development of emphysematous changes, depending on the animal species. In some species, such as mice, cigarette smoke-induced neutrophil inflammation and lung injury resolve after cigarette smoke exposure is discontinued [40, 41], which differs from the pathogenesis of COPD in humans. Additionally, lipopolysaccharide is used in animal models of COPD-like pathogenesis, which induces airway and parenchymal changes. However, the pathophysiology of lipopolysaccharide-induced animal models is more similar to acute respiratory distress syndrome than that of COPD. Although these stimulants are sometimes combined to create models that mimic the pathophysiology of COPD or COPD exacerbations, capturing the full pathophysiology of COPD remains challenging [42].

4.3.2 Differences Between Mouse and Human AT2 Cells

The process of alveolar epithelial cell regeneration in injured lungs also differs between mice and humans. In mouse lungs, bronchioalveolar stem cells (BASCs), marked by co-expression of club cell marker Scgb1a1 (also called CC10) and AT2 cell marker Sftpc, have been reported to contribute to the repair of alveolar epithelial cells after the lung injury, whereas the presence of BASCs has not been confirmed in human lung [43].

4.3.3 Differences Between Human Pluripotent Stem Cell-Derived and Primary AT2 Cells

In recent years, AT2 cells derived from pluripotent stem cells (PSCs)—such as human embryonic stem cells and induced PSCs (iPSCs)—have become a powerful tool in human cell research [44, 45]. PSCs can provide an adequate supply of AT2 cells for research owing to their unrestricted proliferative capacity. Advances in culture methods have enabled the long-term culture of PSC-derived AT2 cells in 3D culture, as well as their successful cryopreservation [46, 47]. Researchers have harnessed PSCs derived from patients with Hermansky-Pudlak syndrome, a rare autosomal recessive hereditary disease that causes pulmonary fibrosis. These PSCs allow the study of AT2 cells with a fibrotic phenotype [48]. In addition, the ease of gene editing has facilitated investigations into the role of specific genes in fibrosis. For instance, editing the causative gene of Hermansky-Pudlak syndrome resulted in the formation of fibrotic AT2 cells [49, 50]. PSC-derived AT2 cells are useful in human pulmonary disease research; however, they have some unique disadvantages compared to primary lung-derived AT2 cells. Since PSC-derived AT2 cells have higher proliferative capacity and are easier to culture than primary AT2 cells, PSC-derived AT2 cells may be more undifferentiated than primary AT2 cells. In addition, reprogramming of differentiated cells during iPSC generation involves resetting of epigenetics and subsequent rejuvenation of the cells. Consequently, these factors may limit investigations into the aging component in age-related pulmonary diseases such as IPF.

As mentioned above, in vitro human primary AT2-derived models are necessary to elucidate the pathogenesis of intractable pulmonary diseases, and the recent studies for isolation and culture methods of human primary AT2 cells are described below.

4.4 Human Lung Dissociation and AT2 Cell Isolation

Isolation of AT2 cells from human lungs involves two major steps: (1) dissociation of the lungs to prepare single-cell suspensions and (2) sorting AT2 cells from these single-cell suspensions.

To obtain single-cell suspensions from human lungs, the tissue is usually subjected to a combination of mechanical and enzymatic digestion in an attempt to cleave cell-cell junctions and degrade the extracellular matrix. Various enzymes are used for lung tissue digestion, including collagenase, dispase, trypsin, or elastase sometimes in combination with deoxyribonuclease 1 (DNase) [51]. Collagenase belongs to the matrix metalloproteinase family, which degrades the collagen triple helices in lung tissue. Dispase is a neutral protease with mild proteolytic activity that cleaves an amino acid sequence that is found in high frequency in collagen IV and fibronectin. Elastase is a highly active elastin-decomposing enzyme and degrades amino acids at sites of the elastin molecule [52]. Elastase can be used alone or in combination with other enzymes to reduce the amount of enzyme used for isolating epithelial cells from the lungs [53]. Trypsin is a natural protease synthesized in the digestive system that breaks down cell-cell junctions. However, it can also severely affect cell membrane proteins, potentially impacting experimental results when surface markers are used for cell isolation [54]. Trypsin also causes free DNA-induced aggregation [55]. These digesting enzymes are either used alone or in combination. To prevent cell aggregation during tissue digestion, DNase is also used in combination with other enzymes. DNase prevents cell aggregation by hydrolytically cleaving phosphodiester bonds and degrades free DNA released from dying cells [54]. DNase is broadly classified into two types: DNase-I and DNase-II. DNase-II is not suitable for tissue digestion for single-cell preparation as it plays a role in engulfment-mediated DNA degradation pathways involved in apoptosis. DNase-I is suitable for the preparation of single-cell suspensions because it prevents cell aggregation by degrading free DNA released from dead cells during enzymatic digestion without inducing the apoptotic pathway.

Sorting AT2 cells from these single-cell suspensions from digested lungs involves identifying characteristic markers of AT2 cells. While expression of the surfactant proteins "SFTPB and SFTPC" and the lipid transporter "ABCA3" are indicative of AT2 cells, these characteristic markers are expressed intracellularly, making them unsuitable for use in live cell isolation. A surface marker, HTII-280, has been reported to be useful for isolating live AT2 cells. Anti-HTII-280 antibody has been used in several reports for the isolation of AT2 cells from human lungs by fluorescence-activated cell sorting (FACS) or magnetic-activated cell sorting [4]. HT2-280 expression decreases in cultured PSC-derived AT2 cells [50]. On the other hand, it has been reported that HTII-280 is expressed in human lung-derived AT2 cells even under passaging culture [56].

Other markers useful for AT2 isolation have also been reported. LysoTracker is a fluorescent dye that stains acidic compartments in live cells. It has been previously shown to selectively label lamellar bodies in cultured AT2 cells for cell sorting by FACS [47, 57]. NaPi2b encoded by *SLC34A2* is a multi-transmembrane protein expressed on AT2 cells; the monoclonal antibody, MX35, recognizing NaPi2b's extracellular domain [58], is also used for the isolation of AT2 cells. The antibody against NaPi2b has almost the same ability to label AT2 cells in human lungs as the antibody against HTII-280 [50]. Dendritic cell-lysosomal associated membrane protein (DC-LAMP)/CD208, a member of the lysosomal associated membrane protein family, is also expressed in mouse/human AT2 cells; furthermore, the antibody against DC-LAMP was used for isolating AT2 cells combined with anti-epithelial cell adhesion molecule/CD326-antibody [59, 60]. These labeling reagents can be used individually or in combination to isolate pure AT2 cells from dissociated human lungs.

4.5 Human Lung-Derived AT2 Cell Culture

AT2 cells were initially cultured in a two-dimensional (2D) culture system [61]; however, this system is not suitable for expanded culture of AT2 cells as it promotes differentiation into AT1 cells, resulting in the loss of stemness. To overcome these problems, AT2 cells are grown in a 3D culture system by embedding them in an extracellular matrix such as Matrigel [62]. The 3D culture system allows AT2 cells to maintain their stemness and self-renew to form an organoid termed the "alveolosphere." The 3D culture method of AT2 cells includes co-culture systems and feeder-free systems. In the co-culture system, mesenchymal cells such as lung fibroblasts are often used as feeder cells. The feeder cells contribute to the proliferation of cultured AT2 cells by mimicking the niche signal required for AT2 cell self-renewal in vivo. For example, PDGFRa$^+$ fibroblasts regulate the self-renewal potential of cultured AT2 cells by secreting wingless related-integration site (Wnt) signaling molecules [6]. Feeder cells are useful for in vitro expansion of AT2 cells but have disadvantages compared to the feeder-free method. When considering cell transplantation after expansion of AT2 cells, there is concern about contamination of feeder cells. Fibroblasts used as feeder cells secrete not only niche-signaling molecules that stimulate AT2 cell proliferation, but also fibrosis-promoting cytokines such as TGF-β, which may be detrimental to the cell-based treatment of some pulmonary diseases including IPF. In addition, it is difficult to evaluate the molecular mechanisms of AT2 cells in a simple system of AT2 cells alone because of the influence of their interactions with feeder cells. Recently, several methods of culturing AT2 cells without feeder cells have been reported [56, 63–68]. By investigating the interaction between lung fibroblasts and lung epithelial cells using single-cell or bulk RNA sequencing, lung fibroblast-derived molecules required for AT2 cell growth were identified [56, 69]. The addition of these identified fibroblast-derived niche factors plus several reagents used in other organ-derived organoid cultures to the medium allowed AT2 cells to be cultured for long-term periods without feeder cells.

4.5.1 Wnt Signaling for AT2 Cell Culture

Wnt signaling pathway plays a crucial role in various biological processes, including normal development, stem cell self-renewal, tissue homeostasis, and regeneration after injury in many organs [70, 71]. Wnt signaling pathway is mainly categorized as canonical (β-catenin dependent) and non-canonical (β-catenin independent). In canonical WNT signaling, the signaling cascade is initiated by the binding of Wnt ligands to the seven-pass transmembrane Frizzled receptor family (FZD) and the low-density lipoprotein receptor-related protein family (LRP5/6), following the nuclear translocation of β-catenin and activation of target genes, including c-Myc and Axin2, via T-cell factor (TCF)/lymphoid enhancer factor (LEF) transcription factors [72]. Non-canonical Wnt signaling involves two pathways: the planar cell polarity (PCP) pathway and the Wnt/Ca^{2+} pathway. The Wnt/PCP pathway activates the JNK and Rho-kinase cascades; the Wnt/Ca^{2+} pathway increases intracellular calcium concentrations and activates protein kinase C, calcineurin, and calcium/calmodulin-dependent protein kinase II cascades [72]. Wnt signaling, especially canonical Wnt signaling, is required for alveolar morphogenesis and AT2 cell proliferation [73]. Axin2-positive AT2 cells, part of the mouse lung AT2 cell population, are tissue stem cells that proliferate in response to Wnt signaling [74, 75]. Human TM4SF1-positive AT2 cells resemble these mouse Axin2-positive cells [75]. In almost all reported methods of human AT2 cell culture without feeder cells, Wnt signal enhancers were used to culture human AT2 cells.

CHIR 99021, a highly selective glycogen synthase kinase 3 (GSK-3) inhibitor [76], is the most frequently used Wnt signal enhancer for AT2 cell culture [56, 64, 66–68]. GSK-3β phosphorylates β-catenin, thereby inhibiting its accumulation in the cytoplasm and concomitant translocation into the nucleus [77]. CHIR 99021 allows β-catenin to translocate into the nucleus by inhibiting GSK-3, thus enhancing Wnt/β-catenin signaling. Ebisudani et al. successfully formed AT2 cell organoids with the addition of Afamin-Wnt-3A to enhance Wnt signaling, whereas the addition of Wnt-3A failed to form AT2 cell organoids [63]. One possible reason for the difference in results is that Afamin-Wnt-3A has a stronger Wnt signaling-enhancing effect than Wnt-3A [78].

Moreover, R-spondins can be added to the medium to enhance Wnt signaling. R-spondins are cysteine-rich glycoproteins that bind to leucine-rich repeat-containing G-protein coupled receptors (LGR) 4/5/6 and E3 ubiquitin ligases ring finger protein (RNF)43/ZNRF3 via their furin-like domains. The R-spondin protein family comprises four members with similar domains (R-spodin 1–4) [79]. R-spondins strongly potentiate canonical Wnt signaling, and their potentiating effects differ among the R-spondin family [80]. R-spondins have been reported to have important roles in lung development [81–83] and injured lung repair [84, 85]. In several reports, R-spondin 1 or 3 was added to the culture medium and contributed to the increase in human AT2 cell organoids [63, 65, 68].

In summary, understanding Wnt signaling pathways provides valuable insights into lung biology and potential therapeutic strategies for pulmonary diseases.

4.5.2 Other Important Culture Media Additives

Several other additives are required to culture human AT2 cells. TGF-β-inducible signaling has been reported to regulate AT2 cell proliferation and differentiation into AT1 cells [86–89]. Bone morphogenetic protein's (BMP) signaling pathway is also involved in AT2 cell proliferation and differentiation into AT1 cells [87, 90, 91]. TGF-β-induced signaling inhibitor, "SB431542 or A8301" [92], and BMP-induced signaling inhibitor, "Noggin" [93], were added to the culture medium in several reports of feeder-free AT2 cell culture [63–66, 68]. In addition, several growth factors and cytokines were sometimes supplemented in the medium to assist AT2 proliferation.

4.6 Conclusion

In recent years, remarkable advancements in research have made it possible to achieve long-term culture of human lung-derived AT2 cells without relying on feeder cells—a previously daunting challenge. In vitro studies using AT2 cells serve as simplified pulmonary disease models and hold great promise in elucidating the pathogenesis of intractable pulmonary diseases.

References

1. Nikolić MZ, Sun D, Rawlins EL. Human lung development: recent progress and new challenges. Development. 2018;145(16):dev163485. https://doi.org/10.1242/dev.163485.
2. Davis JD, Wypych TP. Cellular and functional heterogeneity of the airway epithelium. Mucosal Immunol. 2021;14(5):978–90. https://doi.org/10.1038/s41385-020-00370-7.
3. Bustamante-Marin XM, Ostrowski LE. Cilia and mucociliary clearance. Cold Spring Harb Perspect Biol. 2017;9(4):a028241. https://doi.org/10.1101/cshperspect.a028241.
4. Evans KV, Lee JH. Alveolar wars: the rise of in vitro models to understand human lung alveolar maintenance, regeneration, and disease. Stem Cells Transl Med. 2020;9(8):867–81. https://doi.org/10.1002/sctm.19-0433.
5. Yasutomo K. Genetics and animal models of familial pulmonary fibrosis. Int Immunol. 2021;33(12):653–7. https://doi.org/10.1093/intimm/dxab026.
6. Barkauskas CE, Cronce MJ, Rackley CR, Bowie EJ, Keene DR, Stripp BR, et al. Type 2 alveolar cells are stem cells in adult lung. J Clin Invest. 2013;123(7):3025–36. https://doi.org/10.1172/jci68782.
7. Tanabe I, Ishimori K, Ishikawa S. Development of an in vitro human alveolar epithelial air-liquid interface model using a small molecule inhibitor cocktail. BMC Mol Cell Biol. 2024;25(1):9. https://doi.org/10.1186/s12860-024-00507-7.
8. Mossel EC, Wang J, Jeffers S, Edeen KE, Wang S, Cosgrove GP, et al. SARS-CoV replicates in primary human alveolar type II cell cultures but not in type I-like cells. Virology. 2008;372(1):127–35. https://doi.org/10.1016/j.virol.2007.09.045.
9. Qian Z, Travanty EA, Oko L, Edeen K, Berglund A, Wang J, et al. Innate immune response of human alveolar type II cells infected with severe acute respiratory syndrome-coronavirus. Am J Respir Cell Mol Biol. 2013;48(6):742–8. https://doi.org/10.1165/rcmb.2012-0339OC.
10. Lindell KO, Liang Z, Hoffman LA, Rosenzweig MQ, Saul MI, Pilewski JM, et al. Palliative care and location of death in decedents with idiopathic pulmonary fibrosis. Chest. 2015;147(2):423–9. https://doi.org/10.1378/chest.14-1127.
11. Katzen J, Beers MF. Contributions of alveolar epithelial cell quality control to pulmonary fibrosis. J Clin Invest. 2020;130(10):5088–99. https://doi.org/10.1172/jci139519.
12. Katzen J, Wagner BD, Venosa A, Kopp M, Tomer Y, Russo SJ, et al. An SFTPC BRICHOS mutant links epithelial ER stress and spontaneous lung fibrosis. JCI Insight. 2019;4(6):e126125. https://doi.org/10.1172/jci.insight.126125.
13. Snetselaar R, van Batenburg AA, van Oosterhout MFM, Kazemier KM, Roothaan SM, Peeters T, et al. Short telomere length in IPF lung associates with fibrotic lesions and predicts survival. PLoS One. 2017;12(12):e0189467. https://doi.org/10.1371/journal.pone.0189467.
14. Kuwano K, Kunitake R, Kawasaki M, Nomoto Y, Hagimoto N, Nakanishi Y, et al. P21Waf1/Cip1/Sdi1 and p53 expression in association with DNA strand breaks in idiopathic pulmonary fibrosis. Am J Respir Crit Care Med. 1996;154(2 Pt 1):477–83. https://doi.org/10.1164/ajrccm.154.2.8756825.
15. Schafer MJ, White TA, Iijima K, Haak AJ, Ligresti G, Atkinson EJ, et al. Cellular senescence mediates fibrotic pulmonary disease. Nat Commun. 2017;8:14532. https://doi.org/10.1038/ncomms14532.
16. Arish N, Petukhov D, Wallach-Dayan SB. The role of telomerase and telomeres in interstitial lung diseases: from molecules to clinical implications. Int J Mol Sci. 2019;20(12):2996. https://doi.org/10.3390/ijms20122996.

17. Barbas-Filho JV, Ferreira MA, Sesso A, Kairalla RA, Carvalho CR, Capelozzi VL. Evidence of type II pneumocyte apoptosis in the pathogenesis of idiopathic pulmonary fibrosis (IFP)/usual interstitial pneumonia (UIP). J Clin Pathol. 2001;54(2):132–8. https://doi.org/10.1136/jcp.54.2.132.
18. Mora AL, Bueno M, Rojas M. Mitochondria in the spotlight of aging and idiopathic pulmonary fibrosis. J Clin Invest. 2017;127(2):405–14. https://doi.org/10.1172/jci87440.
19. Enomoto Y, Katsura H, Fujimura T, Ogata A, Baba S, Yamaoka A, et al. Autocrine TGF-β-positive feedback in profibrotic AT2-lineage cells plays a crucial role in non-inflammatory lung fibrogenesis. Nat Commun. 2023;14(1):4956. https://doi.org/10.1038/s41467-023-40617-y.
20. Selman M, Pardo A. The leading role of epithelial cells in the pathogenesis of idiopathic pulmonary fibrosis. Cell Signal. 2020;66:109482. https://doi.org/10.1016/j.cellsig.2019.109482.
21. Yao L, Conforti F, Hill C, Bell J, Drawater L, Li J, et al. Paracrine signalling during ZEB1-mediated epithelial-mesenchymal transition augments local myofibroblast differentiation in lung fibrosis. Cell Death Differ. 2019;26(5):943–57. https://doi.org/10.1038/s41418-018-0175-7.
22. Yao L, Zhou Y, Li J, Wickens L, Conforti F, Rattu A, et al. Bidirectional epithelial-mesenchymal crosstalk provides self-sustaining profibrotic signals in pulmonary fibrosis. J Biol Chem. 2021;297(3):101096. https://doi.org/10.1016/j.jbc.2021.101096.
23. Adeloye D, Chua S, Lee C, Basquill C, Papana A, Theodoratou E, et al. Global and regional estimates of COPD prevalence: systematic review and meta-analysis. J Glob Health. 2015;5(2):020415. https://doi.org/10.7189/jogh.05.020415.
24. Adeloye D, Song P, Zhu Y, Campbell H, Sheikh A, Rudan I. Global, regional, and national prevalence of, and risk factors for, chronic obstructive pulmonary disease (COPD) in 2019: a systematic review and modelling analysis. Lancet Respir Med. 2022;10(5):447–58. https://doi.org/10.1016/s2213-2600(21)00511-7.
25. Syamlal G, Kurth LM, Dodd KE, Blackley DJ, Hall NB, Mazurek JM. Chronic obstructive pulmonary disease mortality by industry and occupation—United States, 2020. MMWR Morb Mortal Wkly Rep. 2022;71(49):1550–4. https://doi.org/10.15585/mmwr.mm7149a3.
26. Wirtz HR, Schmidt M. Acute influence of cigarette smoke on secretion of pulmonary surfactant in rat alveolar type II cells in culture. Eur Respir J. 1996;9(1):24–32. https://doi.org/10.1183/09031936.96.09010024.
27. Xue Y, Williams TL, Li T, Umbehr J, Fang L, Wang W, et al. Type II pneumocytes modulate surfactant production in response to cigarette smoke constituents: restoration by vitamins A and E. Toxicol In Vitro. 2005;19(8):1061–9. https://doi.org/10.1016/j.tiv.2005.05.003.
28. Sims MW, Tal-Singer RM, Kierstein S, Musani AI, Beers MF, Panettieri RA, et al. Chronic obstructive pulmonary disease and inhaled steroids alter surfactant protein D (SP-D) levels: a cross-sectional study. Respir Res. 2008;9(1):13. https://doi.org/10.1186/1465-9921-9-13.
29. Wert SE, Yoshida M, LeVine AM, Ikegami M, Jones T, Ross GF, et al. Increased metalloproteinase activity, oxidant production, and emphysema in surfactant protein D gene-inactivated mice. Proc Natl Acad Sci U S A. 2000;97(11):5972–7. https://doi.org/10.1073/pnas.100448997.
30. Yoshida M, Whitsett JA. Alveolar macrophages and emphysema in surfactant protein-D-deficient mice. Respirology. 2006;11(Suppl):S37–40. https://doi.org/10.1111/j.1440-1843.2006.00806.x.
31. Gamble E, Grootendorst DC, Hattotuwa K, O'Shaughnessy T, Ram FS, Qiu Y, et al. Airway mucosal inflammation in COPD is similar in smokers and ex-smokers: a pooled analysis. Eur Respir J. 2007;30(3):467–71. https://doi.org/10.1183/09031936.00013006.
32. Shapiro SD. End-stage chronic obstructive pulmonary disease: the cigarette is burned out but inflammation rages on. Am J Respir Crit Care Med. 2001;164(3):339–40. https://doi.org/10.1164/ajrccm.164.3.2105072c.
33. Cox RA, Burke AS, Jacob S, Oliveras G, Murakami K, Shimoda K, et al. Activated nuclear factor kappa B and airway inflammation after smoke inhalation and burn injury in sheep. J Burn Care Res. 2009;30(3):489–98. https://doi.org/10.1097/BCR.0b013e3181a28e13.

34. Watanabe N, Fujita Y, Nakayama J, Mori Y, Kadota T, Hayashi Y, et al. Anomalous epithelial variations and ectopic inflammatory response in chronic obstructive pulmonary disease. Am J Respir Cell Mol Biol. 2022;67(6):708–19. https://doi.org/10.1165/rcmb.2021-0555OC.
35. Miles T, Hoyne GF, Knight DA, Fear MW, Mutsaers SE, Prêle CM. The contribution of animal models to understanding the role of the immune system in human idiopathic pulmonary fibrosis. Clin Transl Immunol. 2020;9(7):e1153. https://doi.org/10.1002/cti2.1153.
36. Moore BB, Lawson WE, Oury TD, Sisson TH, Raghavendran K, Hogaboam CM. Animal models of fibrotic lung disease. Am J Respir Cell Mol Biol. 2013;49(2):167–79. https://doi.org/10.1165/rcmb.2013-0094TR.
37. Tanner L, Single AB. Animal models reflecting chronic obstructive pulmonary disease and related respiratory disorders: translating pre-clinical data into clinical relevance. J Innate Immun. 2020;12(3):203–25. https://doi.org/10.1159/000502489.
38. Sawada M, Ohno Y, La BL, Funaguchi N, Asai T, Yuhgetsu H, et al. The Fas/Fas-ligand pathway does not mediate the apoptosis in elastase-induced emphysema in mice. Exp Lung Res. 2007;33(6):277–88. https://doi.org/10.1080/01902140701509458.
39. Takahashi A, Majumdar A, Parameswaran H, Bartolák-Suki E, Suki B. Proteoglycans maintain lung stability in an elastase-treated mouse model of emphysema. Am J Respir Cell Mol Biol. 2014;51(1):26–33. https://doi.org/10.1165/rcmb.2013-0179OC.
40. Groneberg DA, Chung KF. Models of chronic obstructive pulmonary disease. Respir Res. 2004;5(1):18. https://doi.org/10.1186/1465-9921-5-18.
41. Williams K, Roman J. Studying human respiratory disease in animals—role of induced and naturally occurring models. J Pathol. 2016;238(2):220–32. https://doi.org/10.1002/path.4658.
42. Ghorani V, Boskabady MH, Khazdair MR, Kianmeher M. Experimental animal models for COPD: a methodological review. Tob Induc Dis. 2017;15:25. https://doi.org/10.1186/s12971-017-0130-2.
43. Basil MC, Morrisey EE. BASC-ing in the glow: bronchioalveolar stem cells get their place in the lung. EMBO J. 2019;38(12):e102344. https://doi.org/10.15252/embj.2019102344.
44. Dye BR, Hill DR, Ferguson MA, Tsai YH, Nagy MS, Dyal R, et al. In vitro generation of human pluripotent stem cell derived lung organoids. Elife. 2015;4:e05098. https://doi.org/10.7554/eLife.05098.
45. Gotoh S, Ito I, Nagasaki T, Yamamoto Y, Konishi S, Korogi Y, et al. Generation of alveolar epithelial spheroids via isolated progenitor cells from human pluripotent stem cells. Stem Cell Rep. 2014;3(3):394–403. https://doi.org/10.1016/j.stemcr.2014.07.005.
46. Jacob A, Morley M, Hawkins F, McCauley KB, Jean JC, Heins H, et al. Differentiation of human pluripotent stem cells into functional lung alveolar epithelial cells. Cell Stem Cell. 2017;21(4):472–88.e10. https://doi.org/10.1016/j.stem.2017.08.014.
47. Yamamoto Y, Gotoh S, Korogi Y, Seki M, Konishi S, Ikeo S, et al. Long-term expansion of alveolar stem cells derived from human iPS cells in organoids. Nat Methods. 2017;14(11):1097–106. https://doi.org/10.1038/nmeth.4448.
48. Strikoudis A, Cieślak A, Loffredo L, Chen YW, Patel N, Saqi A, et al. Modeling of fibrotic lung disease using 3D organoids derived from human pluripotent stem cells. Cell Rep. 2019;27(12):3709–23.e5. https://doi.org/10.1016/j.celrep.2019.05.077.
49. Chen YW, Huang SX, de Carvalho A, Ho SH, Islam MN, Volpi S, et al. A three-dimensional model of human lung development and disease from pluripotent stem cells. Nat Cell Biol. 2017;19(5):542–9. https://doi.org/10.1038/ncb3510.
50. Korogi Y, Gotoh S, Ikeo S, Yamamoto Y, Sone N, Tamai K, et al. In vitro disease modeling of Hermansky-Pudlak syndrome type 2 using human induced pluripotent stem cell-derived alveolar organoids. Stem Cell Rep. 2019;12(3):431–40. https://doi.org/10.1016/j.stemcr.2019.01.014.
51. Bondonese A, Craig A, Fan L, Valenzi E, Bain W, Lafyatis R, et al. Impact of enzymatic digestion on single cell suspension yield from peripheral human lung tissue. Cytometry A. 2023;103(10):777–85. https://doi.org/10.1002/cyto.a.24777.

52. Yuan H, Kononov S, Cavalcante FS, Lutchen KR, Ingenito EP, Suki B. Effects of collagenase and elastase on the mechanical properties of lung tissue strips. J Appl Physiol. 2000;89(1):3–14. https://doi.org/10.1152/jappl.2000.89.1.3.
53. Mao P, Wu S, Li J, Fu W, He W, Liu X, et al. Human alveolar epithelial type II cells in primary culture. Physiol Rep. 2015;3(2):e12288. https://doi.org/10.14814/phy2.12288.
54. Reichard A, Asosingh K. Best practices for preparing a single cell suspension from solid tissues for flow cytometry. Cytometry A. 2019;95(2):219–26. https://doi.org/10.1002/cyto.a.23690.
55. Panchision DM, Chen HL, Pistollato F, Papini D, Ni HT, Hawley TS. Optimized flow cytometric analysis of central nervous system tissue reveals novel functional relationships among cells expressing CD133, CD15, and CD24. Stem Cells. 2007;25(6):1560–70. https://doi.org/10.1634/stemcells.2006-0260.
56. Katsura H, Sontake V, Tata A, Kobayashi Y, Edwards CE, Heaton BE, et al. Human lung stem cell-based alveolospheres provide insights into SARS-CoV-2-mediated interferon responses and pneumocyte dysfunction. Cell Stem Cell. 2020;27(6):890–904.e8. https://doi.org/10.1016/j.stem.2020.10.005.
57. Van der Velden JL, Bertoncello I, McQualter JL. LysoTracker is a marker of differentiated alveolar type II cells. Respir Res. 2013;14(1):123. https://doi.org/10.1186/1465-9921-14-123.
58. Yin BW, Kiyamova R, Chua R, Caballero OL, Gout I, Gryshkova V, et al. Monoclonal antibody MX35 detects the membrane transporter NaPi2b (SLC34A2) in human carcinomas. Cancer Immun. 2008;8:3.
59. Abdelwahab EMM, Rapp J, Feller D, Csongei V, Pal S, Bartis D, et al. Wnt signaling regulates trans-differentiation of stem cell like type 2 alveolar epithelial cells to type 1 epithelial cells. Respir Res. 2019;20(1):204. https://doi.org/10.1186/s12931-019-1176-x.
60. Salaun B, de Saint-Vis B, Pacheco N, Pacheco Y, Riesler A, Isaac S, et al. CD208/dendritic cell-lysosomal associated membrane protein is a marker of normal and transformed type II pneumocytes. Am J Pathol. 2004;164(3):861–71. https://doi.org/10.1016/s0002-9440(10)63174-4.
61. Jain KG, Xi NM, Zhao R, Ahmad W, Ali G, Ji HL. Alveolar type 2 epithelial cell organoids: focus on culture methods. Biomedicine. 2023;11(11):3034. https://doi.org/10.3390/biomedicines11113034.
62. Barkauskas CE, Chung MI, Fioret B, Gao X, Katsura H, Hogan BL. Lung organoids: current uses and future promise. Development. 2017;144(6):986–97. https://doi.org/10.1242/dev.140103.
63. Ebisudani T, Sugimoto S, Haga K, Mitsuishi A, Takai-Todaka R, Fujii M, et al. Direct derivation of human alveolospheres for SARS-CoV-2 infection modeling and drug screening. Cell Rep. 2021;35(10):109218. https://doi.org/10.1016/j.celrep.2021.109218.
64. Konishi S, Tata A, Tata PR. Defined conditions for long-term expansion of murine and human alveolar epithelial stem cells in three-dimensional cultures. STAR Protoc. 2022;3(2):101447. https://doi.org/10.1016/j.xpro.2022.101447.
65. Salahudeen AA, Choi SS, Rustagi A, Zhu J, van Unen V, de la O SM, et al. Progenitor identification and SARS-CoV-2 infection in human distal lung organoids. Nature. 2020;588(7839):670–5. https://doi.org/10.1038/s41586-020-3014-1.
66. Shiraishi K, Nakajima T, Shichino S, Deshimaru S, Matsushima K, Ueha S. In vitro expansion of endogenous human alveolar epithelial type II cells in fibroblast-free spheroid culture. Biochem Biophys Res Commun. 2019;515(4):579–85. https://doi.org/10.1016/j.bbrc.2019.05.187.
67. van Riet S, van Schadewijk A, Khedoe P, Limpens R, Bárcena M, Stolk J, et al. Organoid-based expansion of patient-derived primary alveolar type 2 cells for establishment of alveolus epithelial Lung-Chip cultures. Am J Physiol Lung Cell Mol Physiol. 2022;322(4):L526–l38. https://doi.org/10.1152/ajplung.00153.2021.
68. Youk J, Kim T, Evans KV, Jeong YI, Hur Y, Hong SP, et al. Three-dimensional human alveolar stem cell culture models reveal infection response to SARS-CoV-2. Cell Stem Cell. 2020;27(6):905–19.e10. https://doi.org/10.1016/j.stem.2020.10.004.

69. Shiraishi K, Shichino S, Ueha S, Nakajima T, Hashimoto S, Yamazaki S, et al. Mesenchymal-epithelial interactome analysis reveals essential factors required for fibroblast-free alveolosphere formation. iScience. 2019;11:318–33. https://doi.org/10.1016/j.isci.2018.12.022.
70. Stabler CT, Morrisey EE. Developmental pathways in lung regeneration. Cell Tissue Res. 2017;367(3):677–85. https://doi.org/10.1007/s00441-016-2537-0.
71. Wu X, van Dijk EM, Ng-Blichfeldt JP, Bos IST, Ciminieri C, Königshoff M, et al. Mesenchymal WNT-5A/5B signaling represses lung alveolar epithelial progenitors. Cells. 2019;8(10). https://doi.org/10.3390/cells8101147.
72. Raslan AA, Yoon JK. WNT signaling in lung repair and regeneration. Mol Cells. 2020;43(9):774–83. https://doi.org/10.14348/molcells.2020.0059.
73. Pongracz JE, Stockley RA. Wnt signalling in lung development and diseases. Respir Res. 2006;7(1):15. https://doi.org/10.1186/1465-9921-7-15.
74. Nabhan AN, Brownfield DG, Harbury PB, Krasnow MA, Desai TJ. Single-cell Wnt signaling niches maintain stemness of alveolar type 2 cells. Science. 2018;359(6380):1118–23. https://doi.org/10.1126/science.aam6603.
75. Zacharias WJ, Frank DB, Zepp JA, Morley MP, Alkhaleel FA, Kong J, et al. Regeneration of the lung alveolus by an evolutionarily conserved epithelial progenitor. Nature. 2018;555(7695):251–5. https://doi.org/10.1038/nature25786.
76. Ring DB, Johnson KW, Henriksen EJ, Nuss JM, Goff D, Kinnick TR, et al. Selective glycogen synthase kinase 3 inhibitors potentiate insulin activation of glucose transport and utilization in vitro and in vivo. Diabetes. 2003;52(3):588–95. https://doi.org/10.2337/diabetes.52.3.588.
77. Rim EY, Clevers H, Nusse R. The Wnt pathway: from signaling mechanisms to synthetic modulators. Annu Rev Biochem. 2022;91:571–98. https://doi.org/10.1146/annurev-biochem-040320-103615.
78. Mihara E, Hirai H, Yamamoto H, Tamura-Kawakami K, Matano M, Kikuchi A, et al. Active and water-soluble form of lipidated Wnt protein is maintained by a serum glycoprotein afamin/α-albumin. Elife. 2016;5:e11621. https://doi.org/10.7554/eLife.11621.
79. Kim KA, Zhao J, Andarmani S, Kakitani M, Oshima T, Binnerts ME, et al. R-spondin proteins: a novel link to beta-catenin activation. Cell Cycle. 2006;5(1):23–6. https://doi.org/10.4161/cc.5.1.2305.
80. Park S, Cui J, Yu W, Wu L, Carmon KS, Liu QJ. Differential activities and mechanisms of the four R-spondins in potentiating Wnt/β-catenin signaling. J Biol Chem. 2018;293(25):9759–69. https://doi.org/10.1074/jbc.RA118.002743.
81. Bell SM, Schreiner CM, Wert SE, Mucenski ML, Scott WJ, Whitsett JA. R-spondin 2 is required for normal laryngeal-tracheal, lung and limb morphogenesis. Development. 2008;135(6):1049–58. https://doi.org/10.1242/dev.013359.
82. Hein RFC, Wu JH, Holloway EM, Frum T, Conchola AS, Tsai YH, et al. R-SPONDIN2(+) mesenchymal cells form the bud tip progenitor niche during human lung development. Dev Cell. 2022;57(13):1598–614.e8. https://doi.org/10.1016/j.devcel.2022.05.010.
83. Szenker-Ravi E, Altunoglu U, Leushacke M, Bosso-Lefèvre C, Khatoo M, Thi Tran H, et al. RSPO2 inhibition of RNF43 and ZNRF3 governs limb development independently of LGR4/5/6. Nature. 2018;557(7706):564–9. https://doi.org/10.1038/s41586-018-0118-y.
84. Raslan AA, Oh YJ, Jin YR, Yoon JK. R-spondin2, a positive canonical WNT signaling regulator, controls the expansion and differentiation of distal lung epithelial stem/progenitor cells in mice. Int J Mol Sci. 2022;23(6):3089. https://doi.org/10.3390/ijms23063089.
85. Zhou B, Magana L, Hong Z, Huang LS, Chakraborty S, Tsukasaki Y, et al. The angiocrine Rspondin3 instructs interstitial macrophage transition via metabolic-epigenetic reprogramming and resolves inflammatory injury. Nat Immunol. 2020;21(11):1430–43. https://doi.org/10.1038/s41590-020-0764-8.
86. Bhaskaran M, Kolliputi N, Wang Y, Gou D, Chintagari NR, Liu L. Trans-differentiation of alveolar epithelial type II cells to type I cells involves autocrine signaling by transforming growth factor beta 1 through the Smad pathway. J Biol Chem. 2007;282(6):3968–76. https://doi.org/10.1074/jbc.M609060200.

87. Chung MI, Bujnis M, Barkauskas CE, Kobayashi Y, Hogan BLM. Niche-mediated BMP/SMAD signaling regulates lung alveolar stem cell proliferation and differentiation. Development. 2018;145(9):dev163014. https://doi.org/10.1242/dev.163014.
88. Riemondy KA, Jansing NL, Jiang P, Redente EF, Gillen AE, Fu R, et al. Single cell RNA sequencing identifies TGFβ as a key regenerative cue following LPS-induced lung injury. JCI Insight. 2019;5(8):e123637. https://doi.org/10.1172/jci.insight.123637.
89. Zhao L, Yee M, O'Reilly MA. Transdifferentiation of alveolar epithelial type II to type I cells is controlled by opposing TGF-β and BMP signaling. Am J Physiol Lung Cell Mol Physiol. 2013;305(6):L409–18. https://doi.org/10.1152/ajplung.00032.2013.
90. Yanagihara T, Zhou Q, Tsubouchi K, Revill S, Ayoub A, Gholiof M, et al. Intrinsic BMP inhibitor Gremlin regulates alveolar epithelial type II cell proliferation and differentiation. Biochem Biophys Res Commun. 2023;656:53–62. https://doi.org/10.1016/j.bbrc.2023.03.020.
91. Zheng X, Chen X, Hu X, Chen L, Mi N, Zhong Q, et al. Downregulated BMP-Smad1/5/8 signaling causes emphysema via dysfunction of alveolar type II epithelial cells. J Pathol. 2024;262(3):320–33. https://doi.org/10.1002/path.6234.
92. Ogunjimi AA, Zeqiraj E, Ceccarelli DF, Sicheri F, Wrana JL, David L. Structural basis for specificity of TGFβ family receptor small molecule inhibitors. Cell Signal. 2012;24(2):476–83. https://doi.org/10.1016/j.cellsig.2011.09.027.
93. Smith WC, Harland RM. Expression cloning of noggin, a new dorsalizing factor localized to the Spemann organizer in Xenopus embryos. Cell. 1992;70(5):829–40. https://doi.org/10.1016/0092-8674(92)90316-5.

Chapter 5
Prevention of Misidentification of Cell Lines/Authentication Methods for Cell Lines

Arihiro Kohara

Abstract Cell lines have become an essential research tool in life science studies and are used by researchers in a variety of fields. However, a certain proportion of cell lines are mislabeled or replaced by cells derived from different individuals, tissues, or species. Misidentification of cell lines in the publication of scientific papers has been overlooked for many years. As a result, thousands of misleading and potentially erroneous papers have been published using misidentified cell lines. Today, methods have been established to authenticate cell lines of human origin, and it is possible to determine if a cell is misidentified. However, still about 8% of human-derived cells deposited in cell banks are found to be misidentified. The problem of misidentification of cell lines is deep-rooted and can only be eradicated through the continuous efforts of all scientists with the same awareness. In this chapter, we make recommendations to scientists and the research community to make cell line authentication a standard practice in laboratories.

Keywords Misidentification · Authentication · STR analysis · Cross-contamination · Cellosaurus

5.1 Introduction

The misidentification of human and animal cell cultures was recognized in the 1950s and continues to be the most serious quality control problem facing the scientific community today [1]. Various methods were developed to solve this problem, with karyotyping and immunological approaches being the first to be used to authenticate cell lines [2–5]. In 1966, Stanley Gertler introduced the concept of

A. Kohara (✉)
JCRB Cell Bank, National Institutes of Biomedical Innovation, Health and Nutrition, Ibaraki, Osaka, Japan
e-mail: kohara@nibn.go.jp

N. Shinomiya, Y. Shimada (eds.), *Applications of Cell Culture*, Current Human Cell Research and Applications, https://doi.org/10.1007/978-981-95-0587-6_5

biochemical polymorphisms to distinguish human cell lines based on isozyme expression and performed the analysis. As a result, Gertler reported that 18 human cell lines supposedly of independent origins were all HeLa cells [6], the first human cancer cell line to be established in culture [7].

The HeLa cell line was derived from the glandular cervical cancer in a female patient named Henrietta Lacks and, because of its celebrated status, was distributed internationally and passed from laboratory to laboratory. Then, as today, many scientists were oblivious to the possibility of cross-contamination. HeLa cells are particularly robust and fast-growing and can proliferate beyond other cells. These included cells claimed to be derived from normal intestinal epithelium (Int-407), normal amnion (WISH), normal liver (Chang liver), laryngeal cancer (Hep-2), and oral cancer (KB).

Many methods have since been used, including human leukocyte antigen (HLA) typing, immunophenotyping, and DNA fingerprinting [8]. Although these methods can authenticate cell lines, they differ in their level of ambiguity and discriminatory power (Table 5.1). The STR profiling method is now used in many laboratories. This method is used by forensic analysts and is also used in criminal investigations, such as DNA identification of criminals (See A for details.).

Table 5.1 Attributes of human cell line identification technologies

Technology	Common application	Means of discrimination	Discriminating power	Limitations
G-banding	Medical diagnosis, research	Chromosomal markers	1.7×10^{5a}	Time/expertise
Immunological	Research	Human cell antigens	None[b]	Specific reagents
Isoenzyme analysis	Speciation of animal cell lines	G6PD polymorphism	2 (types A and B)[c]	Limited discrimination
HLA typing	Medical diagnosis, research	HLA type, subtype	155,448 alleles[d]	Time/expertise
DNA fingerprinting	Identity to the individual level	Hypervariable genomic content		Time/interspecies detection
STR profiling 6 loci 16 loci	Identity to the individual level	Hypervariable genomic content	Locus dependent $0.1–3.5 \times 10^7$ $1.3–5.0 \times 10^{18}$	Interspecies detection

[a] Discrimination is based upon the presence within some human cell lines of specific chromosomal markers. This theoretically determined value may be far in excess of the number observed empirically

[b] This is dependent upon the specific antiserum used. As typically performed, antiserum specific to a given species antigen (e.g., nucleoprotein) has been employed, resulting in the ability to discriminate only to the species level

[c] A polymorphism for the enzyme glucose 6-phosphate dehydrogenase leading to two types (types A and B) has been used to discriminate among human cell lines

[d] As of 10 July 2019. Reference: New Allele Frequency Database: https://www.allelefrequencies.net/default.asp

Once the authentication method was established by STR analysis, leading cell banks around the world introduced it as one method of quality control for human cells, began validating cell lines registered in their banks, and are now authenticating all human cells they distribute. Additionally, the number of misidentified cells found in the current cell bank was reported to be 14–18% in 1999 [9, 10]. Japanese cell banks (JCRB and RCB) also reported in 2014 that misidentified cell lines represented less than 10% of deposited cells [11]. An example of a misidentified cell with a larger impact is the report by [12] that one of NCI-60 distributed for the evaluation of drugs and other cancer-related studies was not a drug-resistant breast cancer cell line (as listed) but an ovarian cancer cell line [12]. Thus, there are currently 545 registered cells that have been misidentified by cell banks and papers [13].

Cell misidentification has many causes and every laboratory is at risk for it. Perhaps the most immediate reason is mislabeling cell culture vessels or freezing vials in routine operations. These errors can be caused by operator workload, lack of attention, or distractions during cell line manipulation. Cross-contamination of cultures and subsequent overgrowth of contaminating cell types are also frequent causes of cell line misidentification. As a way to prevent such cross-contamination and misidentification, the American Society for Cell Biology (ASCB) has published nine guidelines (Table 5.2) [14]. The content is very basic, but the main points are well summarized and should be observed by those responsible for educational positions to their staff and students. Sharing of reagents, repeated use of the same pipette during resupply operations, and simultaneous manipulation of multiple cultures in the same safety cabinet increase the risk of cross-contamination and misidentification.

Table 5.2 Guidelines to prevent cross-contamination and misidentification

No.	Contents
1	If possible, do cell culture work at a time of day when distractions and fatigue will be minimal
2	Before and after working with cultures, thoroughly wipe the hood with a disinfectant
3	Work with only one cell line or cell lineage in the hood at one time
4	Disinfect the hood prior to and after each cell line is fed, subcultured, or handled for other purposes
5	Each cell line or lineage must be fed, rinsed, or trypsinized, with reagents specifically dedicated for that line. In other words, never share reagents among different cell lines, even if the cells are maintained in the same medium formulation. Each cell line must have a dedicated set of reagents and supplies
6	Plan ahead to ensure that you have an adequate supply of reagents and other materials. If you find yourself short you may be tempted to break the cardinal rule: NEVER SHARE REAGENTS
7	A pipette should be used only once, regardless of the number of flasks to be fed. Fill and dispense only once. Never go back into a bottle of reagent with a used pipette
8	Examine and evaluate the cultures before they are brought to the hood. Check the cultures for growth and morphological characteristics
9	Whenever possible, use authenticated cell lines obtained from a repository with high standards for certification and quality control

Furthermore, a survey that profiled active cell culture workers found that of 483 respondents, 32% used HeLa cells, 9% unwittingly were using HeLa contaminants, only 33% of the investigators tested their cell lines for authenticity, and 35% obtained their cell lines from other laboratories rather than from a major repository [15]. Under the current circumstances, it is easy to see the importance of educating and raising the awareness of researchers involved in culture.

To put an end to this problem, the American Type Culture Collection (ATCC) Standards Development Organization (SDO) Workgroup ASN-0002 was established in 2007. This workgroup developed best practices (standards) for use in the life sciences and published guidelines [16] that balance the perspectives of industry, government, regulatory agencies, and academia. This work has led to revisions in the rules for submitting papers and in the peer review process, and many scientific journals now require that the authentication information of the cell lines used in a paper must accompany the paper when it is published.

5.2 Case Study

Since the establishment of highly accurate testing methods such as molecular biological methods, cell misidentification has been investigated in cell banks around the world, and many misidentified cells have been reported. Among the misidentified cells reported, some of the most controversial ones are introduced below.

5.2.1 Misidentification by T-24 (Bladder Cancer Cell Line)

T-24 is a very fast proliferating cell and has been found in many cell misidentifications. In 1999, ECV304 cells (originally described as spontaneously immortalized normal endothelial cells) were shown to be T24 cells [17]. However, since the publication of this paper, many articles have continued to describe ECV304 cells as endothelial cells. For example, within 10 years of its publication, we can find papers treating ECV304 as a vascular endothelial cell in well-known journals such as Nature Immunol. 6(5), 497 (2005) and Nature Biotechnol. 25, 921 (2007). Furthermore, there was no immediate decrease in the number of publications using ECV304. More than 20 years after its publication, there are still papers using ECV304 as a vascular endothelium in 2023, although the number of papers is much smaller now [18].

In Cancer Research in 2001, it was shown that TSU-Pr1 are T24 bladder cancer cells [19]. In the same journal, less than 3 years later, TSU-Pr1 cells were used as a prostate cancer model (Cancer Res. 64, 1058–1066 (2004)). TSU-Pr1 are continued to be used in some studies as a model for prostate cancer, such as Endocrinology 147, 530–542 (2006) and Cancer Cell 5, 67 (2004).

5.2.2 *Misidentification by HeLa (Cervical Cancer Cell Line)*

HSG are cells established as salivary gland carcinoma-derived cells [20]. However, when the cells were deposited in the JCRB cell bank in 2003, they were mistakenly identified as HeLa cells, which was subsequently published in a paper [21]. A survey was conducted in Japan to determine if the original cells existed, but could not find any HSG that differed from HeLa. The HSG cells were used to analyze the in vitro pharmacological effects of drugs for the treatment of Sjögren's syndrome and were included in the medical package insert. In other words, this is an example of misinformation being used to evaluate a drug using cells (HeLa) that are not salivary gland carcinoma cells.

HEp-2 is a cell distributed by the WHO for polio vaccine testing [22]. This line was originally thought to be derived from an epidermoid carcinoma of the larynx [23, 24] but was subsequently found—based on isoenzyme analysis, HeLa marker chromosomes, and DNA fingerprinting—to have been established via HeLa cell contamination. However, because of its susceptibility to poliovirus, it is still used as a cell line for polio vaccine testing, even though it is known to be a HeLa derivative.

5.2.3 *Misidentification by HEK293 (Transformed Embryonal Kidney Cell Line)*

Invossa-K was the first gene therapy drug approved by MFSD in Korea in 2017 [25]. Invossa-K is a treatment for osteoarthritis, and allogeneic chondrocytes transfected with TGF-β1 were supposed to have been used. However, in fact, the transgenic cells were not chondrocytes, but HEK-293 is injected into patients [26]. HEK-293 cells themselves are tumorigenic [27] and can induce tumors in the body. Although the approval was subsequently withdrawn in 2019, the cells have been injected into more than 3000 patients, and there are concerns about tumor development as a side effect in the future.

5.3 Historical Evolution of Authentication Methods

5.3.1 *Early Cell Line Authentication Methods: Karyotyping and Immunological (Serological) Techniques*

The earliest methods for species-verification included chromosomal analysis (karyotyping) and immunological techniques. Karyotypic analysis using Giemsa and quinacrine staining allows species identification based on chromosome number and chromosome structure characteristics [4, 28, 29]. It is also possible to discriminate between human cells by identifying characteristic chromosomal markers (deletions, insertions, duplications, and reciprocal translocations) [30]. Such chromosomal markers exist for many cell lines derived from many disease states and tumors. The usefulness of karyotype analysis in the differentiation of human cell lines extends primarily to cell lines derived from tumors prone to specific chromosomal rearrangements (hematopoietic, mesenchymal, and nervous system) [31]. The disadvantage of cell line authentication by karyotyping and banding is that it is very labor intensive and requires skilled and experienced analysts. Even if characteristic markers are observed, the method is not accurate enough to determine the identity of individual cells.

Immunological (serological) methods for species-level authentication were also employed in the late 1950s and early 1960s [4]. These methods depend on the development of reagents for species-specific antigens and involve immunofluorescence or hemagglutination endpoints.

5.3.2 Cell Line Authentication Methods in the 1960s: Isozyme Analysis Method

Isozyme analysis began to be employed for confirmation of cell line identity in the late 1960s [32, 33]. Isozyme analysis can be used for both species-level authentication and limited discrimination between human cell lines based on polymorphisms in the enzymes G6PDH and PGM [34–36]. In fact, these polymorphisms enabled Gartler [5] to demonstrate the extent to which HeLa cells had cross-contaminated other human cell lines. Compared to karyotyping, isozyme analysis is relatively rapid and can be performed without specialized skills. Isoenzyme analysis is also capable of detecting interspecies cell line cross-contamination in the range of 5%–11% [37]. The major disadvantage is the limited discrimination power.

5.3.3 Molecular Biological Approach: Cellular Markers

The detection of specific cell markers constitutes one of the molecular approaches for cell authentication. An example of an important marker that may be used to differentiate human cell lines is human leukocyte antigen (HLA typing) [38]. The discriminatory power of this technique is based on the relatively large number of alleles conferring HLA types and subtypes within the human population. HLA typing has better identification power than other techniques that have been used, but not as much power as DNA fingerprinting techniques.

5.3.4 Molecular Biological Approach: DNA Profiling

Analysis of DNA fingerprints obtained using probes for tandem repeat minisatellite sequences within hypervariable loci was first described in 1985 [39]. Fingerprinting techniques were subsequently characterized specifically for the authentication of human cell lines [40]. Although this method can unambiguously identify cell lines, the data generated are not reproducible enough across laboratories to allow this method to be used in standardized reference databases [41]. Short Tandem Repeat (STR) profiling, on the other hand, is a DNA profiling method that examines/screens STRs (DNA elements repeated in tandem, 2–6 base pairs long) in human chromosomes and is not only rapid and cost-effective, but is also a standard reference. It has been shown to produce reproducible data in a format suitable for use in databases [41]. Recently, single nucleotide polymorphism (SNP) panels have become available, and this technique has high discriminatory power [42, 43]. A panel of approximately 48 SNPs will provide discriminatory power equivalent to that obtained with STR kits targeting 15 autosomal STR loci [43, 44]. SNP-based analysis is a relatively new method for human identification and has not been extensively adopted [45–47]. Other methods have been developed to identify individuals by sequencing 200 SNPs at a depth of 3000× by next-generation sequencers [48].

5.4 Authentication of Human Cell Lines: STR Analysis

5.4.1 Overview

One of the most commonly used methods for human cell line identification is STR analysis (DNA profiling) [41] (Fig. 5.1). STR analysis is a common molecular biology method used to compare allele repeats at specific loci in DNA between two or more samples. A short tandem repeat is a microsatellite with repeat units that are 2–7 base pairs in length, with the number of repeats varying among individuals, making STRs effective for human identification purposes [49]. This method differs from restriction fragment length polymorphism analysis (RFLP) since STR analysis does not cut the DNA with restriction enzymes. Instead, polymerase chain reaction (PCR) is employed to discover the lengths of the short tandem repeats based on the length of the PCR product. STR analysis consists of DNA extraction of cells, confirmation of DNA quantity and quality, performing multiplex PCR, identification of STR region repeat counts, and analysis of detection data. STR analysis involves simultaneous amplification of several STR marker (e.g., D5S818, D13S317, D7S820, D16S539, vWA, TH01, TPOX, CSF1PO) and the amelogenin gene for gender determination. For each STR marker used, the power of discrimination improves by about an order of magnitude. Thus with 8 STRs, random match probabilities on the order of 1 in 100 million are expected between cell line DNA samples originating from unrelated individuals. Each unique human cell line has a

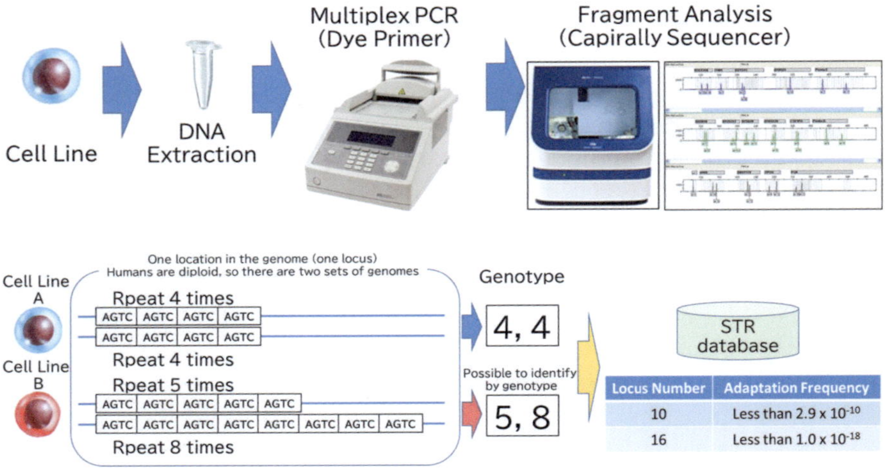

Fig. 5.1 Overview of STR analysis of cell lines

distinct DNA profile, and when the STR DNA fragment sizes are converted to numeric values, the DNA profiles are readily compared among different laboratories.

5.4.2 DNA Extraction of Cell Lines

There are several procedures and technologies available for the isolation of genomic DNA from human cell lines and tissues. In general, all the procedures require (1) lysis of cells by detergent, (2) protein and other contaminant removal by proteinase K treatment, (3) RNA removal, and (4) recovery of pure DNA by precipitation with either ethanol or isopropanol. Commercially available DNA extraction kits can be used instead. Suitable methods vary depending on the type of sample (cell, tissue, fixed specimen, etc.) used for extraction, so a suitable method should be selected.

5.4.3 DNA Quantity and Quality

After DNA extraction, the quantity and quality of the extracted DNA must be confirmed in order to perform PCR in the next step. The most commonly used techniques for measuring DNA concentration are UV absorbance at 260 nm (spectrophotometry) [50–53]. The spectrophotometric assay is based on the amount of UV radiation that is absorbed by the DNA bases. If the sample is essentially free of contaminants such as RNA, proteins, and phenol, this method is accurate and easy to perform. There are other DNA quantification methods using fluorescent reagents (SYBR®Gold [54], PicoGreen® [55], etc.). This method using fluorescent reagents is a highly sensitive method for measuring the amount of reagent intercalated in DNA and is capable of accurately quantifying even small amounts of DNA. The amount of DNA required for a single analysis is sufficient at about 1 ng/10 μL.

The quality of the DNA will also be important for the next step of PCR. In particular, some materials for DNA extraction (e.g., FFPE or DNA stored at room temperature for a long period) may have fragmented DNA, making it difficult to obtain the desired amplified product by PCR. It is necessary to confirm that the extracted DNA is not fragmented using a commercial fragmentation confirmation kit (Genomic DNA ScreenTape [56], FRED Assay Kit [57] etc.).

5.4.4 *Performing Multiplex PCR*

Analysis of length polymorphism at short tandem repeat (STR) loci utilizing multiplex polymerase chain reaction (PCR) remains the primary method for genotyping forensic samples. PCR amplification of specific polymorphic regions (i.e., STR loci) using DNA that has been tested for quantity and quality. Multiplex PCR kits [58] (PowerPlex® 16 STR System [59], Identifiler Plus PCR Amplification Kit [60], Investigator IDplex Plus Kit [61], etc.) available from various companies are used for this PCR (Table 5.3). The Combined DNA Index System (CODIS) is the US National DNA database created and maintained by the Federal Bureau of Investigation. These amplification kits amplify the core CODIS loci [62] evenly distributed on the chromosomes: D3S1358, D5S818, D7S820, D8S1179, D13S317, D16S539, D18S51, D21S11, CSF1PO, FGA, TH01, TPOX, vWA. The main component of the kit is a dye primer with a fluorescent probe for each locus. PCR of DNA extracted from cells with the primers in this kit allows each STR Loci to be amplified with a fluorescent probe attached.

5.4.5 *Capillary Sequencing for Fragment Analysis of Multiplex PCR Amplified Products*

A genetic analyzer (capillary-based sequencer) is used to both separate the post-PCR samples and to collect data on the fragments that include both size and yield. This allows the number of repetitions per STR locus to be determined. Samples migrate through an electrical field; the phosphate groups on the DNA have a negative charge, so the sample migrates towards the anode. Keeping the DNA single-stranded ensures that all same-sized fragments migrate together. Sample aliquots are prepared in a formamide solution, together with an Internal Size Standard (ISS). Water may be used in place of formamide, but samples should not wait for more

Table 5.3 Multiplex PCR kits for authentication of human cell lines by STR profiling

	PowerPlex® 16 STR System	Identifiler Plus PCR Amplification Kit	Investigator IDplex Plus Kit
Maker	Promega	Thermo Fisher Scientific	Qiagen
No. of loci	15 + amelogenin	15 + amelogenin	15 + amelogenin
Marker FBI CODIS Core STR loci	D21S11, CSF1PO, vWA, D8S1179, TH01, D18S51, D5S818, D16S539, D3S1358, TPOX, FGA, D7S820, D13S317, amelogenin	D21S11, CSF1PO, vWA, D8S1179, TH01, D18S51, D5S818, D16S539, D3S1358, TPOX, FGA, D7S820, D13S317, amelogenin	D21S11, CSF1PO, vWA, D8S1179, TH01, D18S51, D5S818, D16S539, D3S1358, TPOX, FGA, D7S820, D13S317, amelogenin
Marker	Penta D, Penta E	D2S1338, D19S433	D2S1338, D19S433

than a few hours for electrophoresis. The samples are denatured into single strands via heat denaturation; the formamide prevents the DNA from denaturing. Single-stranded fragments are passed through a laser, which excites a dye attached to one of the primers in each pair within the multiplex reaction. A detector captures the intensity of the dye, if any single is detected at all.

The ISS has a separate color from any used in the PCR. The ISS has a range of known fragment sizes, which are referenced by the data analysis software to assign a size to each allelic ladder for each locus that can be used as a marker for identifying alleles. An allele ladder is run in a separate lane, and the number of STR locus repeats can be determined by comparing this ladder to the sample peaks.

Capillary sequencers can be used to detect differences in the length of a single nucleotide, so the number of repeats in the STR region with 4–6 base repeats can be well-determined. By adjusting the length of the PCR amplified product through primer design, it is possible to identify the repeat number of 5 Loci at once with a single type of fluorescent probe. By selecting fluorescent probes with different excitation and fluorescence wavelengths such as 6-FAM, LIZ, NED, PET, and VIC, it is possible to analyze more than 20 loci in a single sample.

5.4.6 Data Analysis

After the genetic analyzer is finished with the run, transfer the data to a computer with the analysis software (GeneMapper™ Software 6 [63], GeneMarker [64] etc.). Select analysis settings as determined through the system validation process. For the very first run for the laboratory, use the default settings to gauge the appropriate settings for future runs and then perform the validation. Set the detection threshold for the dye labeling of the ISS higher than for the STR profile loci. This higher threshold enables a higher selection of scoring true ISS peaks instead of any background anomalies. If the ISS dye threshold is too high to score all peaks, lower it accordingly and relabel the ISS. Examine each sample/ladder/control. Be sure that all of the ISS are detected and properly scored. The negative control should have no dye peaks other than the ISS; the positive control should have all alleles detected and properly scored. If the controls do not work as expected, the run is null and void. When evaluating samples, exclude peaks that are clearly artifacts (urea spikes, stutter peaks, etc.) and note this fact. Add calls for any low peaks that appear legitimate yet filtered. Unlike normal diploid material, peaks in cell lines may have varying heights. Do not score any peaks below the analytical threshold because below this baseline level is background.

5.4.7 *Interpretation for STR Profiling Data of Human Sample (Including Cell Lines)*

The result of the sample peaks determined by data analysis is the STR profile of the cell line. The actual interpretation of the results is done by comparing the STR profile of the sample with the database. If the same cell line as the sample is registered in the database, the results of the analysis are verified whether they match the registered data or not. On the other hand, if the same cell line as the sample is not registered in the database (e.g., if a new cell line is established), the results can be verified and determined to be different from the STR profile registered so far in the database. In this way, the STR profile of the cell line (sample) you used is compared with the STR profile registered in the database to perform cell authentication.

When comparing to databases, you can use the Match Criteria Worksheet provided by ICLAC (https://iclac.org/wp-content/uploads/ICLAC_Match-Criteria-Worksheet_v1_3.docx). Cell lines may undergo genetic drift with passage. This can be minimized through good cell culture practice—for example, by returning to early passage frozen stocks rather than passaging a cell line for prolonged periods. However, even with good cell culture practice, genetic drift is known to occur in some cell lines. This may result in differences when comparing STR profiles from the same donor. In 2012, the American Type Culture Collection Standards Development Organization (ATCC SDO) Workgroup ASN-0002 developed a standard for authentication of human cell lines by STR profiling, published by the American National Standards Institute. After revision in 2021, the standard includes detailed protocols for DNA extraction, STR profiling, data analysis, and quality control of the data and provides support in interpreting STR results in case of allelic dropouts by loss of heterozygosity or multi-allelic gains due to microsatellite instability [65]. The Standard recommends a set of match criteria, allowing for some degree of genetic drift when interpreting STR profile results from human cell lines. Recommendations in the Standard are supported by more than 10 years of publications in this area [41, 66]. With the calculation of the percentage of match, the following is done. (1) Obtain an STR Profile for your Test Sample. (2) Identify a Reference Sample for Comparison. (3) Compare Test and Reference Samples using the Match Criteria given. Reference Sample data can be used if the cells are registered in Cellosaurus [67] (https://www.cellosaurus.org/index.html). Data comparison between the Test Sample and the Reference Sample is performed by applying the following formula to find the percentage of match.

$$\text{Match Algorithm} = \frac{\text{SHARED ALLELES} \times 2}{\text{TOTAL ALLELES in the Test Sample} + \text{TOTAL ALLELES in the Reference Sample}}$$

$$\text{Percent Match} = \text{Match Algorithm result} \times 100$$

If there is no reference data, enter the Test Sample data into the databases provided by ATCC, DSMZ, Cellosaurus (CLASTR [68]), etc., and perform a comparison with the registered data. It returns results in order of the highest to lowest percent match.

Interpret the results in three ranges. (A) The range 0–55%: this result is consistent with the two samples being unrelated (different donors). (B) The range 56–79%: this result is indeterminant and may need further testing. (C) The range 80–100%: this result is consistent with the two samples being related (same donor).

5.5 Factors Influencing Result Determination

5.5.1 Overview

Many of the human-derived cell lines that have been established are derived from cancer tissue. Cancer has acquired genomic instability during its proliferative process and has an increasing tendency to cause DNA mutations (alterations) and other genetic changes. This genomic instability can be broadly classified into chromosomal instability (CIN) and microsatellite instability (MSI), both of which are induced by replication stress [69]. MSI is induced in a mismatch repair (MMR)-deficient background and is also induced when DSBs are "misrepaired by microhomology-mediated end-joining repair (MMEJ)." MSI is induced in a mismatch repair (MMR)-deficient background and is induced in a "microhomology-mediated end-junction repair (MMEJ)" background of DSBs that also arise [70]. There are various other factors that can affect the results of STR analysis, such as LOH and Y chromosome dropout.

5.5.2 Effects of CIN

Changes in the number of chromosomes or STR regions amplified or deleted in cancer cell lines due to CIN cause changes in peak heights in STR analysis data. For example, if a locus has different maternal and paternal repeat counts, and the number of chromosomes carrying that locus increases by 1–3, the ratio of the earlier repeat counts would be 1:2.

5.5.3 *Effects of MSI*

Although in vitro evolution of tumor cell lines is well known [71], the underlying genomic alterations often remain obscure. Despite strict quality control through lot-by-lot STR genotyping, applied scientists are unaware of important genetic changes in the tumor models used by the cell lines. If the cell line used is (MMR)-deficient, then the genomic changes based on it are dramatic. As a result, changes in the number of repeats in the STR analysis results accumulate significantly [72]. The previous NGS analyses of cancer tissue have shown that more than 5% of endometrial, stomach, small intestine, colon and rectum, cervix, prostate bile duct and liver adenocarcinomas, neuroendocrine tumors, non-epithelial ovarian cancer, and uterine sarcomas are MMR deficient [73, 74]. Thus, cell lines derived from these cancers are prone to changes in the number of STR repeats. If changes due to genomic instability involving changes in the number of chromosomes occur in MSH-H cancer-derived cell lines, three or more repeat counts will appear at the STR locus.

5.5.4 *Effect of Amelogenin Y Dropout*

Various commercial STR-PCR multiplex reaction kits co-amplify the amelogenin locus [75] in conjunction with autosomal STR loci for gender and individual identification, respectively. The amelogenin locus is a single copy gene with homologs present in the X- and Y-chromosomes (AMGX and AMGY, respectively) differing in both size and sequence [76, 77]. However, it is often observed that the AMGY regions listed in the STR-PCR kit are dropped out during the process of cell establishment and passaging. Therefore, even if a male-derived cell line does not show a peak derived from AMGY, it cannot be immediately ruled out as a cellular misidentification. On the other hand, if AMGY is found in a female-derived cell line, it can be judged as a cell misidentification with a high probability (the exception is XY females).

5.5.5 *Effect of Loss of Heterozygosity (LOH)*

Normal somatic cells have two copies of every autosome and every gene on that chromosome, but cultured cells, tumor tissue samples, and even human pluripotent stem cells (hPSCs) can lose or gain copies of chromosome segments or even the entire chromosomes [77]. For cell lines without high genomic instability, LOH can occur, resulting in convergence to the paternal or maternal chromosome repeat count.

Table 5.4 List of STR search engines

Database	Management	URL
CLASTR	Swiss Institute of Bioinformatics	https://www.cellosaurus.org/str-search/
Search the STR Database	ATCC	https://www.atcc.org/search-str-database
CellDive	DSMZ	https://celldive.dsmz.de/
Cell Authentication Database	JTCA	https://jcrbcelldata.nibiohn.go.jp/str/
STRBase	NIST	https://strbase-archive.nist.gov/

5.5.6 *Database Matching Analysis*

Web-based authentication using STR search engines linked to STR reference databases is indispensable for cell authentication [78]. Most importantly, the Cellosaurus database was launched by Amos Bairoch and his team in 2012. The Cellosaurus is a knowledge resource that aims to document all cell lines used in biomedical research [67] and provides already a wealth of information on more than 113,673 human cell lines (current release May 2024). Moreover, the Cellosaurus warns the user when a cell line is known to be problematic (e.g., misidentified, partly contaminated, misclassified), and assigns a unique Research Resource Identifier (RRID) to all cell lines as it participates in the Resource Identification Initiative [79]. In 2019, the Cellosaurus was expanded by the Cellosaurus STR similarity search tool (CLASTR), which enables researchers to compare their obtained STR profiles with those available in the Cellosaurus database [68]. The current Cellosaurus release contains STR profiles of more than 8000 distinct human cell lines and is by far the largest cell line STR profile database. Taken together, all prerequisites, including a wealth of knowledge, robust authentication methods, comprehensive databases, and tools, that enable researchers to detect misidentified or cross-contaminated cell lines are in place and easily accessible. In addition, ATCC, DSMZ, and the Japanese Society for Tissue Culture prepare and maintain their own STR search engines for public use (Table 5.4).

5.6 Conclusion

Misidentification of cell lines still occurs today and has a significant impact on the research community, including the waste of money, time, and labor. Therefore, the research community must continue to engage in efforts to eliminate this. As an appropriate way to deal with this issue, we introduce the recommendations of Souren et al. [80] Journals and publishers play an important role in preventing the publication of misidentified cell line data. Scientists and research institutions also play an equally important role in preventing the generation of such data by avoiding the use of misidentified cell lines. Experience has shown that regular authentication of cell lines is not common practice in the research community and requires fundamental changes in the preparation of scientific projects. Therefore, we propose the following recommendations for scientists and research institutions to ensure that cell line authentication becomes standard practice in their laboratories:

1. Regular authentication of old and new cell lines.
2. Training in proper cell culture practices.
3. Certification policies by funding agencies and research institutions.

The scientific and research communities should take firm steps now to ensure that cell misidentification does not hinder the future development of science.

References

1. Nardone RM. Curbing rampant cross-contamination and misidentification of cell lines. BioTechniques. 2008;45:221–7.
2. Rothfels KH, Axelrad AA, Siminovitch L, McCulloch EA, Parker RC. In: Begg RW, editor. Proceedings of the third Canadian cancer research conference. New York: Academic Press; 1958. p. 189–214.
3. Defendi V, Billimgham RE, Silvers WK, Moorhead P. Immunological and karyological criteria for identification of cell lines. J Natl Cancer Inst. 1960;25:359–85.
4. Brand KG, Syverton JT. Results of species-specific hemagglutination tests on "transformed", nontransformed, and primary cell cultures. J Natl Cancer Inst. 1962;28:147–57.
5. Gartler SM. Genetic markers as tracers in cell culture. Natl Cancer Inst Monogr. 1967;26:167–95.
6. Nelson-Rees WA, Flandermeyer RR, Hawthorne PK. Banded marker chromosomes as indicators of intraspecies cellular contamination. Science. 1974;184:1093–6.
7. Gey GO, Coffman WD, Kubicek MT. Tissue culture studies of the proliferative capacity of cervical carcinoma and normal epithelium. Cancer Res. 1952;12:264–5.
8. American Type Culture Collection Standards Development Organization Workgroup ASN-0002. Cell line misidentification: the beginning of the end. Nat Rev Cancer. 2010;10:441–8.
9. MacLeod RAF, et al. Widespread intraspecies cross-contamination of human tumor cell lines arising at source. Int J Cancer. 1999;83:555–63.
10. Drexler HG, Dirks WG, MacLeod RA. False human hepatopoietic cell lines: cross-contaminations and misinterpretations. Leukemia. 1999;13:1601–7.
11. Kohara A, et al. Cell misidentification: current status and actions required of researchers. Exp Med. 2014;32:1413–8.
12. Liscovitch M, Ravid D. A case study of misidentification of cancer cell lines: MCF-7/AdR cells (re-designated NCI/ADR/RES) are derived from OVCAR-8 human ovarian carcinoma cells. Cancer Lett. 2006;245(1–2):350–2.
13. Register of misidentified cell lines, International Cell Line Authentication Committee (ICLAC), Version 13 Table 1.
14. Ouellette LA, Nardone RM. Cross-contamination of cell cultures: a call for vigilance and authentication. ASCB Newsletter, July 2007.
15. Buehring GC, Eby EA, Eby MJ. Cell line cross-contamination: how aware are mammalian cell culturists of the problem and how to monitor it? In Vitro Cell Dev Biol Anim. 2004;40:211–5.

16. Authentication of human cell lines: standardization of short tandem repeat (STR) profiling—revised 2022, ANSI/ATCC ASN-0002-2022.
17. Dirks WG, Drexler HG, MacLeod RAF. ECV304 (endothelial) is really T24 (bladder carcinoma): cell line cross-contamination at source. In Vitro Cell Dev Biol. 1999;35:558–9.
18. Guo W, Huang D, Li S. Lycopene alleviates oxidative stress-induced cell injury in human vascular endothelial cells by encouraging the SIRT1/Nrf2/HO-1 pathway. Clin Exp Hypertens. 2023;45(1):2205051.
19. van Bokhoven A, Varella-Garcia M, Korch C, Miller GJ. TSU-Pr1 and JCA-1 cells are derivatives of T24 bladder carcinoma cells and are not of prostatic origin. Cancer Res. 2001;61(17):6340–4.
20. Shirasuna K, Sato M, Miyazaki T. A neoplastic epithelial duct cell line established from an irradiated human salivary gland. Cancer. 1981;48:745–52.
21. Lin LC, Elkashty O, Ramamoorthi M, Trinh N, Liu Y, Sunavala-Dossabhoy G, Pranzatelli T, Michael DG, Chivasso C, Perret J, Chiorini JA, Delporte C, Tran SD. Cross-contamination of the human salivary gland HSG cell line with HeLa cells: a STR analysis study. Oral Dis. 2018;24(8):1477–83.
22. Black FL, Melnick JL, Reissig M. Propagation of measles virus in a strain of human epidermoid cancer cells (Hep-2). Proc Soc Exp Biol Med. 1956;93(1):107–8.
23. Toolan HW. Transplantable human neoplasms maintained in cortisone-treated laboratory animals: H.S. No. 1; H.Ep. No. 1; H.Ep. No. 2; H.Ep. No. 3; and H.Emb.Rh. No. 1. Cancer Res. 1954;14(9):660–6.
24. Moore AE, Sabachewsky L, Toolan HW. Culture characteristics of four permanent lines of human cancer cells. Cancer Res. 1955;15(9):598–602.
25. Evans C, Ghivizzani S, Robbins P. Arthritis gene therapy approved in Korea. J Am Acad Orthop Surg. 2018;26:e36–8.
26. Mobasheri A. Future cell and gene therapy for osteoarthritis (OA): potential for using mammalian protein production platforms, irradiated and transfected protein packaging cell lines for over-production of therapeutic proteins and growth factors. In: Turksen K, editor. Cell biology and translational medicine, volume 8. Advances in experimental medicine and biology, vol. 1247. Cham: Springer; 2019.
27. Shen C, Gu M, Song C, Miao L, Hu L, Liang D, Zheng C. The tumorigenicity diversification in human embryonic kidney 293 cell line cultured in vitro. Biologicals. 2008;36(4):263–8.
28. Rothfels KH, Aelrad AA, Siminovitch L, McCulloch EA, Parker RC. In: Begg RW, editor. Proceedings of the 3rd Canadian cancer conference. New York: Academic Press; 1959. p. 189–214.
29. Defendi V, Billingham RE, Silvers WK, Moorhead P. Immunological and karyological criteria for identification of cell lines. J Natl Cancer Inst. 1960;25:359–85.
30. An International System for Human Cytogenetic Nomenclature. Report of the Standing Committee on Human Cytogenetic Nomenclature. The National Foundation March of Dimes, 1978.
31. MacLeod RA, Drexler HG. Cytogenetic characterization of tumor cell lines. Methods Mol Med. 2004;88:57–76.
32. Gartler SM. Apparent Hela cell contamination of human heteroploid cell lines. Nature. 1968;217(5130):750–1.
33. de Oca FM, Macy ML, Shannon JE. Isoenzyme characterization of animal cell cultures. Proc Soc Exp Biol Med. 1969;132(2):462–9.
34. O'Brien SU, Kleiner G, Olson R, Shannon JE. Enzyme polymorphisms as genetic signatures in human cell cultures. Science. 1977;195(4284):1345–8.
35. O'Brien SJ, Shannon JE, Gail MH. A molecular approach to the identification and individualization of human and animal cells in culture: isozyme and allozyme genetic signatures. In Vitro. 1980;16(2):119–35.
36. Wright WC, Daniels WP, Fogh J. Distinction of seventy-one cultured human tumor cell lines by polymorphic enzyme analysis. J Natl Cancer Inst. 1981;66(2):239–47.

37. Nims RW, Shoemaker AP, Bauernschub MA, Rec LJ, Harbell JW. Sensitivity of isoenzyme analysis for the detection of interspecies cell line cross-contamination. In Vitro Cell Dev Biol Anim. 1998;34(1):35–9.
38. Ferrone S, Pellegrino MA, Reisfeld RA. A rapid method for direct HL-A typing of cultured lymphoid cells. J Immunol (Baltimore, Md.: 1950). 1971;107(2):613–5.
39. Jeffreys AJ, Wilson V, Thein SL. Individual-specific 'fingerprints' of human DNA. Nature. 1985;316(6023):76–9.
40. Gilbert DA, Reid YA, Gail MH, Pee D, White C, Hay RJ, O'Brien SJ. Application of DNA fingerprints for cell-line individualization. Am J Hum Genet. 1990;47(3):499–514.
41. Masters JR, Thomson JA, Daly-Burns B, Reid YA, Dirks WG, Packer P, Toji LH, Ohno T, Tanabe H, Arlett CF, Kelland LR, Harrison M, Virmani A, Ward TH, Ayres KL, Debenham PG. Short tandem repeat profiling provides an international reference standard for human cell lines. Proc Natl Acad Sci U S A. 2001;98(14):8012–7.
42. Kidd KK, Pakstis AJ, Speed WC, Grigorenko EL, Kajuna SL, Karoma NJ, Kungulilo S, Kim JJ, Lu RB, Odunsi A, Okonofua F, Parnas J, Schulz LO, Zhukova OV, Kidd JR. Developing a SNP panel for forensic identification of individuals. Forensic Sci Int. 2006;164(1):20–32.
43. Phillips C, Fang R, Ballard D, Fondevila M, Harrison C, Hyland F, Musgrave-Brown E, Proff C, Ramos-Luis E, Sobrino B, Carracedo A, Furtado MR, D Syndercombe Court, Schneider PM, SNPforID Consortium. Evaluation of the Genplex SNP typing system and a 49plex forensic marker panel. Forensic Sci Int: Genet. 2007;1(2):180–5.
44. Pietrangeli I, Ottaviani E, Martone C, Gabriele L, Arcudi G, Potenza S, Spinella A, Giardina E, Novelli G. Frequency assessment of 25 SNPs in five different populations. Forensic Sci Int Genet. 2010;4(5):e131–3.
45. Castro F, Dirks WG, Fahnrich S, Hotz-Wagenblatt A, Pawlita M, Schmitt M. High-throughput SNP-based authentication of human cell lines. Int J Cancer. 2013;132:308–14.
46. Yu M, Selvaraj SK, Liang-Chu MM, Aghajani S, Busse M, Yuan J, Lee G, Peale F, Klijn C, Bourgon R, et al. A resource for cell line authentication, annotation and quality control. Nature. 2015;520:307–11.
47. Zhang Q, Luo M, Liu CJ, Guo AY. CCLA: an accurate method and web server for cancer cell line authentication using gene expression profiles. Brief Bioinform. 2021;22:bbaa093.
48. Chen X, Qian W, Song Z, Li QX, Guo S. Authentication, characterization and contamination detection of cell lines, xenografts and organoids by barcode deep NGS sequencing. NAR Genom Bioinform. 2020;2(3):lqaa060.
49. Butler JM. Advanced topics in forensic DNA typing: methodology. San Diego: Elsevier Academic Press; 2011. p. 99–100.
50. Warburg O, Christian W. Isolation and crystallization of enolase. Biochem Z. 1942;310:384–421.
51. Gallagher SR. Quantitation of DNA and RNA with absorption and fluorescence spectroscopy. In: Ausubel FM, Brent R, Kingston RE, Moore DD, Seidman JG, Smith JA, Struhl K, editors. Short protocols in molecular biology. New York: Greene Publishing Associates and John Wiley & Sons; 1992. p. A3-5–8.
52. Gallagher SR. Quantitation of DNA and RNA with absorption and fluorescence spectroscopy. In: Ausubel FM, Brent R, Kingston RE, Moore DD, Seidman JG, Smith JA, Struhl K, editors. Current protocols in molecular biology, vol. 3. New York: John Wiley & Sons; 1995. p. A.3D.1–8.
53. Sambrook J, Fritch E, Maniatis T. Molecular cloning: a laboratory manual. In: Section E.5. 2nd ed. Cold Springs Harbor, NY: Cold Spring Harbor Laboratory Press; 1989.
54. Tuma RS, Beaudet MP, Jin X, Jones LJ, Cheung CY, Yue S, Singer VL. Characterization of SYBR gold nucleic acid gel stain: a dye optimized for use with 300-nm ultraviolet transilluminators. Anal Biochem. 1999;268(2):278–88.
55. Ahn SJ, Costa J, Emanuel JR. PicoGreen quantitation of DNA: effective evaluation of samples pre- or post-PCR. Nucleic Acids Res. 1996;24(13):2623–5.
56. Burrell A, Foy C, Burns M. Applicability of three alternative instruments for food authenticity analysis: GMO identification. Biotechnol Res Int. 2011;2011:838232.

57. Mano J, Nishitsuji Y, Kikuchi Y, Fukudome SI, Hayashida T, Kawakami H, Kurimoto Y, Noguchi A, Kondo K, Teshima R, Takabatake R, Kitta K. Quantification of DNA fragmentation in processed foods using real-time PCR. Food Chem. 2017;226:149–55.
58. Butler JM. Genetics and genomics of core STR loci used in human identity testing. J Forensic Sci. 2006;51(2):253–65.
59. Bao B, Ali S, Banerjee S, Wang Z, Logna F, Azmi AS, Kong D, Ahmad A, Li Y, Padhye S, Sarkar FH. Curcumin analogue CDF inhibits pancreatic tumor growth by switching on suppressor microRNAs and attenuating EZH2 expression. Cancer Res. 2012;72(1):335–45.
60. Wang DY, Chang CW, Lagacé RE, Calandro LM, Hennessy LK. Developmental validation of the AmpFℓSTR® Identifiler® Plus PCR Amplification Kit: an established multiplex assay with improved performance. J Forensic Sci. 2012;57(2):453–65.
61. Kraemer M, Prochnow A, Bussmann M, Scherer M, Peist R, Steffen C. Developmental validation of QIAGEN Investigator® 24plex QS Kit and Investigator® 24plex GO! Kit: two 6-dye multiplex assays for the extended CODIS core loci. Forensic Sci Int Genet. 2017;29:9–20.
62. Hares DR. Selection and implementation of expanded CODIS core loci in the United States. Forensic Sci Int Genet. 2015;17:33–4.
63. Chatterji S, Pachter L. Reference based annotation with GeneMapper. Genome Biol. 2006;7(4):R29.
64. Holland MM, Parson W. GeneMarker® HID: a reliable software tool for the analysis of forensic STR data. J Forensic Sci. 2011;56(1):29–35.
65. ANSI/ATCC ASN-0002-2021. Authentication of human cell lines: standardization of Short Tandem Repeat (STR) profiling, 2021.
66. Tanabe H, Takada Y, Minegishi D, Kurematsu M, Msui T, Mizusawa H. Cell line individualization by STR multiplex system in the cell bank found cross-contamination between ECV304 and EJ-1/T24. Tissue Cult Res Commun. 1999;18:329–38.
67. Bairoch A. The Cellosaurus, a cell-line knowledge resource. J Biomol Tech. 2018;29(2):25–38.
68. Robin T, Capes-Davis A, Bairoch A. CLASTR: the Cellosaurus STR similarity search tool—a precious help for cell line authentication. Int J Cancer. 2020;146(5):1299–306.
69. Fujimori H, Hyodo M, Matsuno Y, Shimizu A, Minakawa Y, Atsumi Y, Nakatsu Y, Tsuzuki T, Murakami Y, Yoshioka KI. Mismatch repair dependence of replication stress-associated DSB recognition and repair. Heliyon. 2019;5(12):e03057.
70. Matsuno Y, Atsumi Y, Shimizu A, Katayama K, Fujimori H, Hyodo M, Minakawa Y, Nakatsu Y, Kaneko S, Hamamoto R, Shimamura T, Miyano S, Tsuzuki T, Hanaoka F, Yoshioka KI. Replication stress triggers microsatellite destabilization and hypermutation leading to clonal expansion in vitro. Nat Commun. 2019;10(1):3925.
71. Ben-David U, Siranosian B, Ha G, Tang H, Oren Y, Hinohara K, Strathdee CA, Dempster J, Lyons NJ, Burns R, Nag A, Kugener G, Cimini B, Tsvetkov P, Maruvka YE, O'Rourke R, Garrity A, Tubelli AA, Bandopadhayay P, Tsherniak A, et al. Genetic and transcriptional evolution alters cancer cell line drug response. Nature. 2018;560(7718):325–30.
72. Koblitz J, Dirks WG, Eberth S, Nagel S, Steenpass L, Pommerenke C. DSMZCellDive: diving into high-throughput cell line data. F1000Res. 2022;11:420.
73. Le DT, Durham JN, Smith KN, Wang H, Bartlett BR, Aulakh LK, Lu S, Kemberling H, Wilt C, Luber BS, Wong F, Azad NS, Rucki AA, Laheru D, Donehower R, Zaheer A, Fisher GA, Crocenzi TS, Lee JJ, Greten TF, et al. Mismatch repair deficiency predicts response of solid tumors to PD-1 blockade. Science. 2017;357(6349):409–13.
74. Hause RJ, Pritchard CC, Shendure J, Salipante SJ. Classification and characterization of microsatellite instability across 18 cancer types. Nat Med. 2016;22(11):1342–50.
75. Lau EC, Mohandas TK, Shapiro LJ, Slavkin HC, Snead ML. Human and mouse amelogenin gene loci are on the sex chromosomes. Genomics. 1989;4(2):162–8.
76. Akane A, Shiono H, Matsubara K, Nakahori Y, Seki S, Nagafuchi S, Yamada M, Nakagome Y. Sex identification of forensic specimens by polymerase chain reaction (PCR): two alternative methods. Forensic Sci Int. 1991;49(1):81–8.

77. Ben-David U. Genomic instability, driver genes and cell selection: projections from cancer to stem cells. Biochim Biophys Acta. 2015;1849(4):427–35.
78. Dirks WG, MacLeod RA, Nakamura Y, Kohara A, Reid Y, Milch H, Drexler HG, Mizusawa H. Cell line cross-contamination initiative: an interactive reference database of STR profiles covering common cancer cell lines. Int J Cancer. 2010;126(1):303–4.
79. Bandrowski A, Brush M, Grethe JS, Haendel MA, Kennedy DN, Hill S, Hof PR, Martone ME, Pols M, Tan S, Washington N, Zudilova-Seinstra E, Vasilevsky N. The resource identification initiative: a cultural shift in publishing. F1000Res. 2015;4:134.
80. Souren NY, Fusenig NE, Heck S, Dirks WG, Capes-Davis A, Bianchini F, Plass C. Cell line authentication: a necessity for reproducible biomedical research. EMBO J. 2022;41(14):e111307.

Chapter 6
Current Knowledge of Circulating Tumor Cell Cultures

Yutaka Shimada

Abstract We previously reviewed on circulating tumor cell (CTC) cell lines and the establishment of each CTC cell line under specific conditions. There is currently no standard method for establishing CTC cell lines. With recent advances in CTC cultures, we reexamined culture methods and CTC research and noted some limitations and an essential crux. Importantly, progress in short-term CTC cultures now allows for their application to precision medicine. The next challenge in this field is the establishment of easy, rapid, and inexpensive methods to cultivate CTCs recovered from individual patients.

Keywords CTCs (circulating tumor cells) · Cell line · Long-term culture · Short-term culture

6.1 Introduction

Circulating tumor cells (CTC) are considered to contribute to the distant metastasis of malignant tumors. However, CTCs do not simply travel in the blood. Although most CTCs are not viable, a few are able to differentiate and express specific characteristics, particularly stem-like features. We successfully cultured CTCs for the first time in 1999 and found that a higher colony formation rate in cultures of blood samples was associated with the number of circulating tumor cells [1, 2]. We then shifted our focus to the detection of CTCs, but not their cultivation, for more than 10 years.

Blood analyses provide only a "snapshot" of tumor cell dissemination. In contrast to studies on circulating nucleic acids, exosomes, and other blood biomarkers,

Y. Shimada (✉)
Department of Nanobio Drug Discovery, Graduate School of Pharmaceutical Sciences, Kyoto University, Sakyo-ku, Kyoto, Japan
e-mail: yshimada@pharm.kyoto-u.ac.jp

N. Shinomiya, Y. Shimada (eds.), *Applications of Cell Culture*, Current Human Cell Research and Applications, https://doi.org/10.1007/978-981-95-0587-6_6

analyses of CTCs define their cellular genetics, epigenetics, transcriptomics, and protein profiles. Sequential analyses enable the assessment of dynamic changes in tumor cells from individual patients. Therefore, the characteristics of CTCs have become an important and attractive theme in recent years.

Although CTC cell lines are useful for repeated evaluations of CTC characteristics, prolonged cultures and multiple passages often lead to phenotypes that are no longer representative of the original tumor in terms of cell epigenetics and gene expression. Therefore, short-term cultures of CTCs may be more useful for precision medicine. However, the successful culture of CTC is still challenging.

We previously reviewed the findings on CTC cultures [2]. Further advances and limitations have since been reported. Among the important findings that were reported, we herein randomly selected and described recent advances in CTC cultures. Importantly, there are still no CTC cell lines of hepato-pancreato-biliary cancer; therefore, we briefly discussed hepato-pancreato-biliary cancer.

6.2 CTC Cultures

6.2.1 CTC Cell Lines

Thirty-nine CTC cell lines had been established by 2021 [2]; however, eight were sequentially established from the same patient at different time points [3, 4]. The majority of CTC cell lines have been from small cell lung cancer (SCLC), breast cancer, and colon cancer (Fig. 6.1). Several cell lines have since been added [5–7]. Four CTC cell lines from SCLC were cultured on ultralow attachment plates [5]. Another study described one SCLC CTC cell line that grew as spheroids [6]. Furthermore, one colon CTC cell line was selected using CD44v6 magnetic beads [7]. Therefore, there are currently less than 40 individual CTC cell lines. Interestingly, there are no CTC cell lines from aggressive cancers, such as hepatic cancer, pancreatic cancer, and bile duct cancer. Moreover, two prostate cancers were cultured only as organoids. Most mainstream growth types show floating spheroid-type patterns, and some cell lines are grown under both two- and three- dimensional (2D and 3D) culture conditions. However, few cell lines have been established on 2D cultures to date. Therefore, 2D and 3D culture conditions are both necessary for CTC cultures that are dependent on the characteristics of tumor cells (Table 6.1). The most important issue for establishing CTC cell lines is the large number of CTC needed (more than 300 cells/mL).

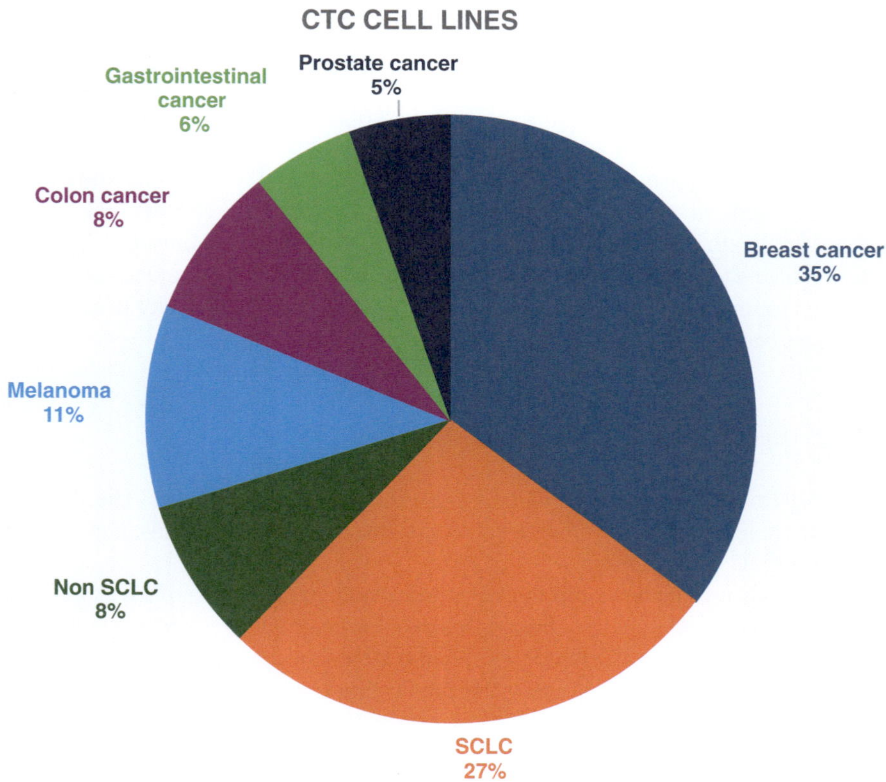

Fig. 6.1 Distribution of CTC cell lines. Most CTC cell lines are from lung cancer (particularly SCLC) and breast cancer

Table 6.1 Culture conditions of CTC cell lines

	2D	3D spheroid	3D organoid	2D + 3D	NA
Breast cancer	4	8		1	
Colon cancer		3			
Small cell lung cancer				5	
Non-small cell lung cancer	1				
Malignant melanoma		3			2
Gastroesophageal cancer				2	
Prostate cancer			2		
Total	5	14	2	8	2

6.2.1.1 Sequential Analysis

Soler et al. [4] derived and characterized nine CTC lines using blood samples from a patient with metastatic colorectal cancer (CRC) collected before and after chemotherapy and targeted therapy as well as during cancer progression. They also described the first permanent cell line (CTC-MCC-41) from circulating colon cancer cells. In CTC-MCC-41.5 line, CTCs started to proliferate in different wells at different time points. They also showed that two spheroids from the same cell line displayed different characteristics [8]. CTCs were present as single cells and clusters in the initial samples and as single cells in the final sample at the time of the third relapse. Moreover, in the first samples, giant and small CTCs were simultaneously detected, whereas only very small (4–6 μm) and small (8 μm) CTCs were present in the final sample before the patient's death. These cell lines displayed an intermediate epithelial/mesenchymal phenotype, stem-cell like characteristics, the potential for angiogenesis, an osteomimetic signature, and the capacity to evade the immune system.

6.2.1.2 Specific Growth Pattern

Some cell lines have been established using ultralow attachment plates. This method forces the selection of CTCs that adapt to specific conditions and, thus, may induce a bias for CTC characteristics.

We already established a cell line of undifferentiated carcinoma of the esophagus (TYUC-1). This cell line has a floating spheroid-type growth pattern with cells that automatically float from the tumor specimen during adhesion cultures [9]. This phenomenon was also observed during the establishment of SCLC CTC cell lines by the Hamilton group and is a natural characteristic of SCLC.

6.2.1.3 Mimicking of the CTC Growth Pattern

Suspension cultures have been shown to mimic the growth pattern of CTCs. Established cell lines that grow as cell clusters or spheroids in suspension may be used as "surrogates" for CTC growth [10, 11].

Park et al. [12] reported the development of a long-term suspension cell culture model using human breast cancer cell lines to mimic CTCs. They found that more than 40 passaged suspension cells produced metastasis-like cancer stem cells. Increases in glutamate and leucine levels and changes in the cholesterol synthesis pathway were observed in suspended cells. Furthermore, cells were unable to proliferate in cell cultures in the absence of glutamine. Therefore, this CTC-mimicking suspension cell culture model may be applied to various cancer types.

Huh et al. [13] recently identified IKZF1, NFE2, BTG2, and IRF8 as adherent-to-suspension transition (AST) factors. AST factors reprogram the anchorage dependency of adherent cells, thereby changing them into suspension cells in an inducible and reversible manner. They demonstrated that blocking the induction of AST factors markedly reduced the number and size of distant lung metastases in a mouse model. These findings expand the prevailing cancer treatment paradigm towards direct interventions for metastatic spread of cancer.

6.2.2 Short-Term Cultures (Within 6 Months)

We previously summarized the methods and culture findings on short-term cultures of CTCs. Many studies have since reported additional and novel information on CTC cultures, several of which are summarized in Table 6.2.

TetherChip technology is a microfluidic device that was engineered to prevent cell adhesion. TetherChip provides a sufficient number of tumor cells for drug evaluations [27]. Moreover, the anchorage of free-floating CTCs in the TetherChip platform allows for the visualization of microtentacles (McTNs), which cannot be observed under adherent conditions. McTNs are tubulin-based protrusions of the plasma membrane that constitute an important mechanism for metastatic dissemination and are also associated with EMT pathways.

Balakrishnan et al. [15] established a unique method using a Matrigel® coating on 96-well plates that enabled cancer cell clusters to attach and proliferate. WBCs did not bind to Matrigel® and were mostly removed during the replenishment of media. They also established a 3D poly ε-caprolactone scaffold system to culture CTCs that imitate the architecture and stiffness of tumors, thereby closely mimicking in vivo tumor growth.

De et al. [16] established a unique ex vivo 3D system (porous poly ε-caprolactone scaffolds) by creating in vivo-like functionality to culture CTCs from whole blood samples without prior enrichment. Proliferating cells included CTCs and other cell types, such as cancer-associated fibroblasts and macrophages, which may provide a suitable microenvironment for the enrichment of CTCs.

A recent study successfully established short-term CTC cultures with viable cells from 5/7 colon, 5/5 lung, and 13/13 pancreatic cancers, resulting in an overall success rate of 92.0% [30].

Although short-term cultures are generally considered to be useful for drug sensitivity evaluations, a drug susceptibility test using CTCs primary cultured in vitro did not establish clinical efficacy in patients with SCLC [18]. This may be attributed to the limited amount of information currently available on the relationship between CTCs and coexisting normal cells.

Furthermore, cultures grown in the presence of CD45+ cells were shown to exhibit higher growth potential ex vivo in breast cancer patients [12, 13]. These findings suggest the importance of leukocytes in CTC cultures.

6.2.2.1 CDX

Xiao et al. [30] established four patient-derived CTC-derived xenograft (CDX) models using a short-term CTC culture (5000–25,000 cells). The gene set that defined epithelial-mesenchymal transition and TGFβ were significantly enriched in all four patients.

Hong et al. [31] established a CDX model using a short-term culture of renal cell cancer (RCC) CTCs with synthetic biological scaffolds. CTC cells were

Table 6.2 Short-term cultures (additional information from a previous review [2])

Research group	Cancer type	Blood volume	CTC enrichment and isolation	Establishment rate or number of patients	Culture conditions	Culture medium	Culture period	Additional information
Arnoletti et al. (2018) [14]	PDAC, bile duct cancer	10–20 mL	FACS, RBC lysis Ficoll-Paque	11 (PDAC), 6 (bile duct)	Co-culture with myeloid-derived fibroblasts	RPMI 1640, 10% medium 199, 10% FCS	7 days	Portal blood, cryogenetically preserved
Balakrishnan et al. (2019) [15]	Breast, lung, esophageal, and bladder cancers	10 mL	RBC lysis	62/151 (cluster)	2D, 1% O_2, ellipsoidal microwell mold in PDMS (polydimethylsiloxane)	DMEM, 10% FCS	21 days	Correlation between cluster formation and treatment response and efficacy
De et al. (2019) [16]	Breast cancer	10 mL	RBC lysis	12/16 (75%)	3D, 1% O_2, 96-well ultra-low attachment plate	DMEM, 10% FCS,	3, 7, 14 days	Polymer scaffold
Lee et al. (2020) [17]	Small cell lung cancer	7.5 mL	Ficoll-Paque, RosetteSep	18/22	2D organoid, binary colloidal crystal (BCC)	Platelet lysate-based culture medium	4 weeks	Binary colloidal crystal (BCC) organoid
Ju et al. (2021) [18]	Small cell lung cancer	7.5 mL	RBC lysis, Ficoll-Paque	5/14	3D, 96-well ultra-low attachment plate	NA	2–4 days	ClearCell FX system, drug sensitivity
Belthier et al. (2021) [7]	Colorectal cancer, breast cancer	6 mL	Ficoll-Paque, RBC lysis, magnetic beads	20/20, one cell line	3D, ultra-low attachment	M12 medium	NA	CD44v6-positive CTCs rarely expressed EpCAM

Lin et al. (2021) [19]	Head and neck cancer	10 mL	Ficoll-Paque, RosetteSep	37/40	2D organoid, binary colloidal crystal (BCC)	Platelet lysate-based culture medium	4 weeks	Binary colloidal crystal (BCC) organoid
Hu et al. (2021) [20]	HCC	NA	Ficoll-Paque	55/60	3D poly-lysine-coated slide	DMEM, Matrigel	12–14 days	Predict outcomes
Wu et al. (2022) [21]	PDAC	20 mL	Ficoll-Paque, RosetteSep	36/41	2D organoid, binary colloidal crystal (BCC)	Platelet lysate-based culture medium	3 weeks	BCC organoid, eSelect system
Arnoletti et al. (2022) [22]	PDAC	10–20 mL	Fluorescent-activated cell sorting	30	Co-culture with myeloid-derived fibroblasts	RPMI 1640, 10% medium 199, 10% FCS	10 days	Portal blood
Pizon et al. (2022) [23]	Breast cancer	NA	Tumorsphere-forming assay	5/10	Co-culture, chorioallantoic membrane (CAM)	RPMI 1640, L-glutamine, EGF, insulin	16 days	Chorioallantoic membrane (CAM)
Morita et al. (2022) [24]	Non-small cell lung cancer	3 mL	Whole blood	2/2	Filter on a 35-mm dish	RPMI 1640, 10% FCS	NA	Soft micropore filter (S-MPF) 10–8 μm
Kwizera et al. (2022) [25]	Breast cancer	1 mL	PEGylated PDMS–titanium–gold electro-microfluidic device (VIZA Chip)	36/60	3D, 24-well ultra-low attachment plate	DMEM/F12, EGF, bFGF, B27, BSA, insulin	60 days	PEGylated PDMS–titanium–gold electro-microfluidic device
Hsieh et al. (2022) [26]	HCC	4 mL	Ficoll-Paque	18/21	Co-culture on mesothelial cell	DMEM 5% FCS	3 days	Drug sensitivity

(continued)

Table 6.2 (continued)

Research group	Cancer type	Blood volume	CTC enrichment and isolation	Establishment rate or number of patients	Culture conditions	Culture medium	Culture period	Additional information
Vardas et al. (2023) [27]	Breast cancer	5 mL	Ficoll-Paque, TetherChip	20/20, expansion to 3–306 cells	3D, microfluidic device	RPMI 1640, 10% FCS, L-glutamine, insulin, transferrin, selenium, EGF, FGF	5 days	Microtentacles (McTNs) were detected
Balakrishnan et al. (2022) [28]	Lung cancer	10 mL	RBC lysis	4/20 (cluster), 1/20 (drug sensitivity)	2D, 1% O_2, ellipsoidal microwells. Matrigel-coated 96-well plate for 2 days	DMEM, 10% FCS	14 days then transfer to a Matrigel-coated plate	Drug sensitivity. WBCs did not bind to Matrigel
Nomura et al. (2023) [29]	Colorectal cancer	10 mL	OncoQuick	18/33	2D, poly (2-methoxyethyl acrylate) (PMEA)-coated plates. 96-well or 24-well Matrigel-coated plate	DMEM/F12, 10% FCS, EGF, HGF, B27, Noggin. R-spondin 1, Y-27632	14 days	Poly (2-methoxyethyl acrylate) (PMEA)-coated plates
Xiao et al. (2023) [30]	Colon, lung, and pancreatic cancers	14 mL	Ficoll-Paque	5/7 (colon cancer), 5/5 (lung cancer), 13/13 (pancreatic cancer), established CDX from 4 selected samples	2D, culture dishes	DMEM/F12 B27, EGF, FGF, heparin, Y-27632, adenine, L-glutamine	59 ± 27 days (lung cancer), 56 ± 25 days (colon cancer), 39 ± 20 days (pancreatic cancer)	CTC expanded in cultures (5000–25,000 cells)

Hong et al. (2023) [31]	Renal cancer	15 mL	Ficoll-Paque	10/15	3D, 96-well ultra-low absorption plate	Stem cell culture	28 days	Biological scaffold hydrogel. Transfer to CDX
Mohamed et al. (2023) [32]	Ovarian cancer, breast cancer	6 mL	Screencell filter	1/7	24-well ultra-low attachment plate	RPMI1640, 20% FCS, Y27532, insulin, FGF, EGF, B-27, cobalt chloride	8 weeks	Cobalt chloride
Furukawa et al. (2023) [33]	Pancreatic cancer	5 mL	Whole blood	15/15	8-μm or 15-μm filter on a 35-mm dish	IMDM-F12, 20% FCS, Matrigel, EGF, bFGF, B27	103 days	Microfilter device
Wang et al. (2023) [34]	Colon, rectal, kidney, and bladder cancers	1.5 mL	PDMS-glass microfluidic device	29	Microfluidic device on a Petri dish	Culture medium, 20% FCS, EGF, FGF	14 days	Functional CTCs
Espejo-Cruz et al. (2023) [35]	HCC	7 mL	Ficoll-Paque	5/22	1% O_2, 6- or 24-well ultra-low attachment plate	DMEM/F12 bFGF, EGF, N2, B27, Y27632	15 days	Hypoxia within 3 days
Che et al. (2023) [36]	Lung cancer	2 mL	Site-specific antibody-conjugated DS platform	109/124	Fibronectin-coated polystyrene culture plate	NA	7 days	Proliferative ability of CTCs in early-stage lung cancer

NA: not available; CTC: circulating cancer cell; CDX: CTC-derived xenograft; SCLC: small cell lung cancer; PDAC: pancreatic ductal adenocarcinoma; CRC: colorectal cancer; 2D: two-dimensional; 3D: three-dimensional

resuspended in stem cell culture medium, which was mixed with synthetic biological scaffolds at a 1:1 volume ratio. Each treated sample was inoculated into a 96-well ultralow absorption plate. Four CDX models were established and used in genetic research.

6.2.2.2 Hepato-Pancreato-Biliary Cancer

Difficulties are associated with establishing hepato-pancreato-biliary cancer CTC cell lines because of the limited number of CTCs detected in these malignancies. Although pancreatic cancer is one of the most aggressive tumor types, the CellSearch® CTC detection method showed that CTC counts were one of the lowest among solid cancers [37]. Nevertheless, short-term cultures of these cancers have been reported, as follows.

Hepatocellular Carcinoma

Espejo-Cruz et al. [35] successfully established CTC cultures from 5 out of 33 patients, mostly using the negative enrichment of CD45(−) cells, ultralow adherence, high glucose, and a short period of hypoxia followed by normoxia. Hypoxia appeared to stimulate the ex vivo proliferation of spheroids from CD45(−)-enriched CTCs from two HCC patients. However, spheroids did not survive long in cultures. The use of magnetic beads coated with anti-EpCAM for the isolation of CTCs with subsequent cultures under ultralow adherence did not provide stable CTC spheres from the peripheral blood of patients with HCC. Furthermore, cell growth was not observed in the IsoFluxTM system. They also noted the presence of many double-positive CD45+ CK+ cells at the time of CTC capture.

Another study recommended 3D (Matrigel) cell cultures [20]. CTC spheroids were detected in 91.67% (55/60) of cultures developed using cells obtained from CTC-positive HCC patients. Spheroids remained healthy until at least day 12–14 of culture.

Short-term co-cultures with human mesothelial cells using a bioreactor chip have also been suggested and are regarded as a useful method for the evaluation of drug sensitivity [26].

Pancreatic Cancer

After the first study on CTC detection by CellSearch®, several methods, such as isolation by the size of epithelial tumor cells (ISET®), MetaCell, Screencell, and NanoVelcro, achieved high detection rates in patients with pancreatic ductal adenocarcinoma (PDAC) [38]. These findings indicated that some pancreatic cancer CTCs do not express EpCAM. Most recently, the RareCyte® platform detected markedly fewer CTCs in pancreatic cancer than in CRC. The RareCyte® platform relies on separation based on the density of CTCs. Therefore, not only EpCAM-dependent but also EpCAM-independent systems showed that pancreatic cancer had the lowest CTC count among the malignancies tested [39].

The collection and analysis of CTCs from the portal vein of PDAC patients is feasible, and the number of CTCs obtained was higher than that from peripheral blood [40–42]. However, it is a more invasive procedure, which limits its applicability.

In a previous study, portal blood samples (10–20 mL) were collected by direct intraoperative venipuncture of the portal vein. After a 7-day ex vivo culture, portal blood mononuclear cells spontaneously formed CTC aggregates with other surrounding cells, which promoted the proliferation of CTCs [14]. In a subsequent study, fluorescent-activated cell sorting-isolated portal PDAC CTCs and myeloid-derived fibroblasts (M-FB) co-cultured ex vivo resulted in greater CTC proliferation, motility, and cluster formation than CTCs cultured alone [22].

The biomimetic cell culture system, eSelect, was expanded ex vivo to form organoids, which were then used to assess sensitivity to a panel of nine drugs. CTC-derived organoid expansion was achieved in 3 weeks, with 87.8% (36/41) culture efficiency [21].

Diagnostic leukapheresis (DLA) has been applied to overcome the low number of CTCs in blood and has significantly improved the detection of CS-CTCs in PDAC. A recent study suggested that a median processed blood volume of 2.8 L was required to generate DLA products with a median volume of 45 mL [43].

A recent study examined circulating stromal cells and cancer-associated macrophage-like cells (CAMLs) in addition to CTCs in PDAC patients using CellSieve, a microfiltration device that isolates cells based on size exclusion. Although 23% of PDAC patients had CTCs, 95% had CAMLs [44]. Furthermore, circulating epithelial cells (CECs) have been detected in premalignant pancreatic lesions, such as intraductal papillary mucinous neoplasm (IPMN) [38].

Biliary Tract Cancer

Short-term cultures of bile duct cancer have been performed using portal blood co-cultured with M-FB [14].

6.3 Capture and Enrichment of CTCs

6.3.1 Identification of CTCs

A definitive method for identifying CTCs has not yet been established. Based on common criteria, EpCAM(+)/DAPI(+)/CD45(−) cells are regarded as CTCs. However, several CTCs do not express EpCAM. Furthermore, some cells express both CK(+) and CD45(+) and are generally used to identify tumor cells. The following markers are also useful: CA9 for RCC, androgen receptor for prostate cancer, ALCAM(CD166) for pancreatic cancer, and ASGPR for hepatic cancer. However, a genetic analysis or historical analysis using xenotransplantation needs to be performed for confirmation.

6.3.2 *Protocols for the Simultaneous Enrichment and Detection of CTCs*

Several methods, such as the Parsortix™ Cell Separation System, Fiber-optic array scanning technology, the CellSearch system, ISET®, AdnaTest, EPISPOT, and the Ariol system, have been employed for the simultaneous enrichment and detection of CTCs.

Wang et al. [34] recently reported that a combination of size-based enrichment and invasiveness-based analyses continuously removed the majority of hemocytes using 8-μm gaps and was useful for assessing the invasiveness of enriched CTCs by Matrigel loading. This system successfully identified functional CTCs derived from patients with different types of cancer, including CRC and kidney and bladder cancers, using whole blood without any sample pretreatment process. In 28 patients with CRC, functional CTCs were detected in 61.54% of those with metastases and exhibited stronger invasiveness evaluated by the distance of migration/invasion than those from patients without metastases ($P < 0.05$).

Cheng et al. [45] described a 3D conductive scaffold microchip that efficiently captured and specifically released CTCs with high purity. Captured CTCs were then electrochemically released by the electro-mediated accurate breakage of the Au–S bond, whereas nonspecifically bonded WBCs were not responsive to this process; therefore, the purity of recovered CTCs was high. The optimized releasing time was set as 5 min.

6.3.3 *Size Selection Method*

6.3.3.1 Points to Consider

Tumor cells do not have a uniform size. One study suggested that CTCs are smaller than previously considered [46]. In breast cancer, the median computed diameter (CD) of patient-derived CTCs was found to be 12.4 μm, while that of cultured cell line cells was 18.4 μm. In prostate cancer, the CD of CTCs was 10.3 μm, while that of cultured cell line cells was 20.7 μm. The CDs of CTCs from CRC and bladder cancer were 7.5 and 8.6 μm, respectively. The CD of leukocytes was 9.4 μm. Therefore, some CTCs are smaller than leukocytes. However, this analysis was limited to EpCAM-positive CTCs.

We remeasured the sizes of cells in CTC cell lines using previously reported values [3, 8, 47] and found that they varied, similar to ordinary cell lines (breast cancer from 12.5 to 21.4 μm and CRC from 7.5 to 20 μm). We employed the filter selection method using CELLNETTA (10-μm pores) and VIVANT CELL POT (8-μm pores) for a pancreatic cancer cell line (Panc-1) and detected cells larger than 10 μm as well as those smaller than 8 μm. Therefore, although 7- or 6-μm pores need to be used to prevent the loss of CTCs, their size varies depending on the cancer type; therefore, it is necessary to determine the correct filter pore size for each cancer type.

SCLC CTCs (approximately 8 μm) are smaller than other tumor CTCs and, thus, may more easily migrate through capillaries and recirculate [48].

6.3.3.2 Brief Description of the Method

Size-dependent CTC enrichments comprise ISET® and diverse filtration devices, such as Metacell, ScreenCellCyto, and the Parsortix™ system [48].

The MetaCell filtration device is a capillary action-driven, size-based separation method that isolates and enriches viable CTCs from peripheral blood samples using a porous polycarbonate membrane. Previous studies demonstrated that CTCs remained viable for 14–28 days by transferring the membrane filter to a cultivation plate and culturing cells directly on the membrane [49–51] and have additionally developed EMT-independent CTC enumeration and harvesting protocols using the Parsortix®.

Several filtration methods have recently been developed. Furukawa et al. [33] reported a microfilter device that uses whole blood to reduce cell damage. Blood samples were passed through a 15-μm pore filter and 8-μm pore filter, and cells were collected by each filter. The filter device was then removed from the syringe and placed on a 35-mm dish, and cells were successfully cultured in all cases (15/15 pancreatic cancer patients). Culture medium flows between the inside and outside of the filter device. CTCs and/or formed clusters and/or colonies were detected during long-term cultures of up to 103 days. In several cases, no CTCs were obtained immediately after filtration; however, single cells, clusters, and colonies were observed after cultivation in each case.

Morita et al. [24] described an affordable and simple filter (hereafter referred to as a soft micropore filter) that comprises a polyethylene film with a thickness of 15 μm and conical pores with a diameter of 8–10 μm, which are uniformly opened (opening rate, 20%). Whole blood is filtered by free-falling under gravity.

Maeshiro et al. [52] reported the CELLNETTA filter. The apparatus is a metal mesh element that has sub-micrometer square apertures arrayed in a periodic manner. Cell aggregates are noninvasively captured by CELLNETTA (mesh sizes of 10, 15, and 20 μm). Medium containing cells is passed through the mesh apparatus into a 50-mL conical tube under the condition that the cell number per passage does not exceed approximately 5×10^5. The mesh apparatus is then inverted over another conical tube, and fresh medium is passed from the back side of the mesh to recover captured aggregates.

6.3.4 *Positive Selection (Improved Method)*

The transmembrane protein, CD44v6, detected CTCs in blood samples from several patients with CRC or breast cancer. CD44v6-positive CTCs rarely express EpCAM, suggesting the need to use a combination of markers for a more comprehensive view of CTC heterogeneity [7]. Using this method, one colon cancer CTC cell line was established.

The VAR2CSA malaria protein efficiently retrieves circulating tumor cells in an EpCAM-independent manner. The malaria rVAR2 protein to oncofetal chondroitin sulfate (ofCS) was shown to efficiently capture CTCs from hepatic, pancreatic, and prostate carcinoma patients with minimal contamination by peripheral blood mononuclear cells. ofCS was expressed by both epithelial and mesenchymal cancer cells and was preserved during the epithelial–mesenchymal transition of cancer cells [53].

6.3.5 *Microbubbles*

Stimulations by surface acoustic waves and bulk acoustic waves have been shown to contribute to the efficient isolation of CTCs from blood samples. By controlling acoustic bubbles, CTCs may be continuously trapped, rotated to form a spheroid, and released into the microchamber for cultivation. A high cell viability (84%) was maintained after 3 days of culture. Acoustic bubbles for spheroid trapping, rotation, and culture: a tumor-on-a-chip platform (ABSTRACT) is capable of forming and culturing viable CTC spheroids. Click bubbles also enable the suspended enrichment of CTCs [54, 55]. Geng et al. [56] described an acoustic-based ultracompact device for cell sorting.

6.4 Culture Medium and Growth Factor

Previous studies suggested that tumor spheroids under a 3D suspension culture were only obtained in a serum-free defined culture medium [25, 47, 57]. However, other studies used 10% FBS [2]. Therefore, this issue may depend on the characteristics of each cell.

EGF, FGF, B27, insulin, Noggin, Y-27639, and TGFb are generally added to cultures. Similar to R1881, which is typically used for prostate cancer, specific factors are applied based on the characteristics of each cancer.

6.5 Culture Conditions

6.5.1 Hypoxia or Normoxia

Hypoxic conditions play an important role in CTC cultures, and hypoxia-inducible factor-1α (HIF-1α) has been identified as a key regulator of this process, activating a range of downstream genes to promote cell proliferation, angiogenesis, and metastasis.

At the time of cluster and spheroid formation, most cells are exposed to hypoxic conditions; however, separate single CTCs are under normoxic conditions. Furthermore, supportive blood cells prefer normoxic conditions and are damaged by hypoxic conditions. Therefore, after the selection of CTCs under hypoxic conditions (typically within 1 week), samples are promptly transferred to normoxic conditions.

Khoo et al. [58] examined three conditions: hypoxia (10% serum, 1% oxygen) in microwells, hypoxia on 2D-uncoated substrates, and normoxia in microwells. Blood cells were present under all three conditions; however, clusters only formed under hypoxia/microwell conditions after 1 week. The percentage of CD44+/CD24− cells was lower on 2D-uncoated substrates or under normoxic conditions than in cultures maintained in microwells under hypoxic conditions. The EMT status of CTCs on Day 14 was heterogeneous, with the majority of cells staining positively for both pan-CK and vimentin antibodies (>50%). Therefore, hypoxia is essential for CTC cultures.

While hypoxic conditions are generally recommended, they are not always critical [59–61].

6.5.2 Ultralow Attachment Plate

Yu et al. [47] previously suggested that CTCs died through senescence after a few cell divisions in monolayer cultures and that growth in suspension and hypoxia were required as implemented in stem cell cultures. Most CTCs were established in a floating environment. However, SCLC CTCs automatically formed spheroids in an ordinary culture plate [48]. The most distinguishing feature of permanent SCLC CTC lines in vitro is the spontaneous formation of large spheroids, termed tumorspheres, a process that has to be enforced by non-adherent cell culture conditions in other cancer cell lines [48].

SCLC CTC lines have been established in 2D cultures under normoxic conditions without enrichment because normal blood cells died in cell cultures and cancer cells persisted. Hypoxic conditions and 3D cultures do not appear to be appropriate for cells in the circulation, but may be suitable for disseminated tumor cells (DTCs). The SCLC CTC panel does not express stem cell antigens or characteristics, and all cells express high levels of EpCAM [5].

Floating cultures using ultralow attachment plates detect specific part of CTC cells. Therefore, this method is not always suitable for research on CTCs.

6.5.2.1 Spheroids

In the eSelect system, cells are seeded onto a binary colloidal crystal substrate containing silica and polymethyl methacrylate and cultured using a platelet lysate-based culture medium. In the case of SCLC, successful CTC spheroid proliferation was established after 4 weeks in 82% of cases (18/22) [17]. The morphological spheroid construct consists of three different types: large-sized, cohesive round-shaped spheroids, small-sized cohesive irregular or round spheroids, and discohesive "grape-like" spheroids. SCLC CTC spheroids have been classified into the following three subpopulations: epithelial, mesenchymal, and intermediate [17].

In the case of head and neck cancer, 37 out of 40 cultures were successfully established, and the ex vivo expansion success rate was 92.50% [19].

6.5.3 Normal Plates

The establishment of adherent 2D culture conditions is technically easier. However, as discussed, CTCs under 2D culture conditions may undergo senescence after several divisions [47]. Nevertheless, several CTC-derived cell lines have been successfully established as adherent.

6.5.4 Co-culture

Several co-cultures with various cell types (tumor-associated white blood cells [62], CD45+ cells [30, 63], and cancer-associated fibroblasts and the extracellular matrix (ECM) [64]) have been shown to support the successful establishment of CTC-derived cultures.

6.5.5 In Ovo *Culture*

One alternative model is an *in ovo* assay using a chorioallantoic membrane (CAM). Primary cultures from circulating cancer stem cells were established using a sphere-forming assay. Tumorspheres were generated from peripheral blood and then transplanted onto the CAM membrane of fertilized chicken eggs to form secondary microtumors. Tumors on CAM were successfully established from 5 out of 10 breast cancer patients within 8 days [23]. Rousset and colleagues generated CAM tumor xenografts from the CTCs of breast cancer, prostate cancer, and lung cancer patients with engraftment efficiencies of 50%, 67%, and 62%, respectively [65].

6.5.6 *Other Novel Methods*

6.5.6.1 DS Platform

The site-specific antibody-conjugated DS platform (abbreviated as "DS platform") originating from 2 mL of blood was seeded onto fibronectin-coated polystyrene culture plates. Lung adenocarcinoma-specific CTCs were enumerated upon isolation and after a 7-day culture. Overall, 88% of captured CTC specimens (109/124) were successfully cultured and qualified for enumeration and analysis. When the median of the cultured number of CTCs, 30, was used as the cutoff, the proliferative ability of CTCs correlated with disease progression [36].

6.5.6.2 Cobalt (II) Chloride

The transition metal cobalt (II) chloride mimics hypoxia by upregulating and stabilizing HIF-1α, which plays an important role in the cell cycle and survival of cancer cells. CTCs were incubated in culture medium supplemented with 100 μM of cobalt (II) chloride as a chemical inducer of HIF-1α and cultured in low adherence plates. Epithelial-like and quasi-mesenchymal CTC phenotypes were isolated from the blood of cancer patients (ovarian and breast) and successfully cultured for more than 8 weeks [32]. Isolated and successfully expanded CTCs expressed EpCAM. The expression of EpCAM by CTCs from hepatocellular carcinoma was found to be upregulated under hypoxia conditions.

6.5.6.3 PEGylated PDMS–Titanium–Gold Electro-Microfluidic Device

Kwizera et al. [25] reported that a robust electro-microfluidic chip efficiently captured heterogeneous (EpCAM+ and CD44+) CTCs with high purity and also induced the glutathione-controlled release of CTCs with high efficiency and viability. Furthermore, CD44+ CTCs were cultured to efficiently form 3D spheroids for scale-up. In contrast, CTCs captured with EpCAM alone did not proliferate well in vitro.

6.5.6.4 Cotton Microfluidic Substrate (CMS)

Arora et al. [66] demonstrated that CMS enabled the static and dynamic isolation of cells from the blood of patients with head and neck squamous cell carcinoma. The CMS matrix exhibits potential for the further development of "tissue on chip" and "point-of-care" medical devices in cancer diagnostics as well as chemotherapeutic efficacy evaluations in both drug discovery and development.

6.5.6.5 Nanoemulsions (NEs)

Another approach is the use of NEs formulated from oleic acids and lipids along with growth factors that stimulate the metabolism of CTCs [67]. A cell suspension was initially seeded in one well of an ultralow attachment 96-well plate and cultured under hypoxic conditions (1–2% O_2) for 1 week. NEs were added to media every 2 days at the same time as fresh medium supplement. After 1 week of culture, cells were switched to standard cell culture conditions. When 85% confluence was reached, growing cells were transferred to an ultralow attachment 24-well plate. This study showed that the gene expression profile of CTCs differed before and after the culture.

6.5.6.6 Poly (2-Methoxyethyl Acrylate) (PMEA)-Coated Plate

Nomura et al. [29] developed a simple method to detect CTCs in the blood of CRC patients using PMEA-coated plates. Blood samples were concentrated by centrifugation in OncoQuick tubes and then incubated overnight on PMEA-coated chamber slides. The next day, cells were transferred to Matrigel-coated 96-well or 24-well plates for subsequent cell culture. Overall, CTCs and/or growing circulating tumor cells were identified in 23/41 CRC cases (56.0%). Most cells showed a spheroid-like morphology. Even among samples in which CTCs were not detected by a cytological survey, the cell culture produced viable spheroid-like structures in 5 of 18 CRC cases.

6.6 DLA

As discussed, DLA is a useful method for enriching CTCs. Leukapheresis is a standard procedure that is routinely used in clinical settings to enrich mononuclear cells (MNCs) from blood for various applications, including stem cell harvest. In DLA, continuous density separation is employed to harvest MNCs, while most of the other blood components, such as plasma, erythrocytes, platelets, neutrophils, basophils, and eosinophils, are returned to the patient. DLA is performed for 1 h.

The use of DLA to isolate CTCs from metastatic breast cancer and metastatic prostate cancer patients significantly increased the yield of CTCs over that with CellSearch® [68]. DLA from 5 L of blood resulted in a median yield of 64 CTCs/mL, in contrast to 2.5 CTCs/mL in peripheral blood samples pre-DLA.

Previous studies demonstrated the feasibility of using DLA for CTC enrichment in patients with advanced breast, prostate, and lung cancers [69–72]. Mout et al. [70] reported that the median yield of CTCs was 5312 with DLA and only 22 in 7.5 mL of blood from patients with metastatic prostate cancer. Organoid expansion from isolated and enriched CTC samples was successful in 35% of cases (14/40). Two organoid cultures (EMC-PCa-25 and EMC-PCa-41) were expanded and maintained for more than 6 months, with the latter yielding a stable cell line.

Franken et al. [73] recovered frozen CTCs from DLA samples. Cryopreserved DLA products were stored with minimal CTC loss and no overt reduction in tumor cell quality or viability during an observation period of up to 3 years. Despite their DLA approach, only CTCs from a few patients were successfully cultured.

6.7 Limitations

6.7.1 Long-Term Culture vs. Short-Term Culture

While short-term cultures may be useful for single-cell analyses and gene expression studies, only long-term cultures generate the number of cells needed to investigate the mechanisms underlying metastasis. Moreover, they have potential as a tool for personalized medicine.

However, epigenetics and gene expression characteristics may be lost in long-term cultures, which involve multiple cell passages, as is the case for CTC lines [58], and, most importantly, cultures may not reflect the heterogeneity observed in patients.

6.7.2 Coexistence of Normal Cells

CTC clusters comprise 2–50 tumor cells, fibroblasts, endothelial cells, and leukocytes with improved survival in the circulation in addition to a higher metastatic potential [48]. Some CTCs interact with platelets to form a shell, which most likely protects them from shear forces in the circulation. Furthermore, CTCs have been assigned hybrid epithelial/mesenchymal phenotypes and stem cell-like properties. SCLC CTCs have been shown to recruit macrophages.

Platelet interactions with tumor cells through either a direct interaction, the secretion of platelet microvesicles, or the release of platelet granules have been implicated in resistance to apoptosis. A potential mechanism for the platelet-induced enhancement of metastasis is based on the secretion of TGF-β from platelets and direct cell surface interactions with CTCs.

6.7.3 CTC-Like Cells Are Present in Healthy Individuals

Captured cells are enumerated after sequential staining for specific markers to identify CTCs, circulating mesenchymal cells (CMCs), putative circulating stem cells (CSCs), and circulating endothelial cells (CECs). The mean cell sizes of CMCs, CECs, and CSCs are approximately 10 μM in diameter (range, 5–15 μM). In clinical studies, CTCs and CMCs were detected in 35% and 58% of cancer patients, respectively, and were largely absent from healthy controls (3%, $p = 0.001$). In contrast, CECs were observed in both cancer and control groups [74].

In patients with chronic obstructive pulmonary disease, circulating epithelial cells were detected 1–4 year prior to lung nodules being observed on CT [75].

Patients with benign inflammatory bowel diseases have been shown to harbor large numbers of epithelial cells in their blood, which cannot always be discriminated from CTCs [76]. In patients with benign colon diseases, positive events that met the criteria for "tumor cells" were detected by both the CellSearch system (11.3%) and the CK19-EPISPOT assay (18.9%), whereas no positive events were observed in samples from healthy volunteers. Positive events were most frequently identified in patients with diverticulosis and Crohn's disease. The CellSearch system detected CK-positive cells in the blood of 6 (11.3%) out of 53 patients, while the CK19-EPISPOT assay revealed CK19-releasing cells in 10 (18.9%). Interestingly, only one sample was positive in both assays.

To avoid contamination by normal epithelial cells, blood obtained immediately after puncture needs to be discarded.

6.7.4 Source of Blood Samples

Uveal melanoma CTCs were detected at a higher frequency and in larger numbers in arterial blood specimens than in venous samples [77]. Similar findings were obtained for stage IV breast cancer and hepatic cancer [78, 79].

In CRC, the liver serves as a filter for CTCs, which are shed in high numbers from the primary tumor. Furthermore, the number of CRC CTCs was markedly higher in the hepatic portosystemic circulation than in the peripheral systemic circulation, indicating that the majority of CTCs from hepatic metastases are trapped during the lung circulation [80]. Previous studies also suggested that CTCs were found at a higher rate and count in the mesenteric venous blood compartment than in the central venous blood compartment [81, 82].

6.7.5 Timing of Sampling

Blood samples collected during the active (10:00 am) and rest (4:00 am) phases of the same day were analyzed. The majority of spontaneous CTC intravasation events occur during sleep. Systemically, key circadian rhythm hormones, such as melatonin, testosterone, and glucocorticoids, dictate CTC generation dynamics, and as a consequence, insulin directly promotes tumor cell proliferation in vivo, but in a time-dependent manner [83].

6.8 Lymphatic CTCs

Experimental evidence to suggest that metastatic cells traverse the lymphatic system prior to distant metastasis is limited. The extent to which lymphatic spread contributes to distant hematogenous spread or whether it is simply a marker of more invasive disease remains unclear [84].

Lymph fluid CTCs (L-CTCs) have recently been attracting increasing attention. They have been shown to contribute to lymph node metastasis and ultimately flow into the bloodstream. Therefore, L-CTCs will inevitably become a subject of CTC research. However, L-CTC cultures are more challenging than CTCs in blood because of the difficulties associated with sampling an adequate amount of lymph fluid. Lymph sampling is not routinely performed in clinical practice. There were studies which used L-CTC source cell lines, named HAS/c and Sum/c [1, 85, 86]. These cell lines were established from rat-planted tumors that originated from thoracic duct fluid (20–30 mL) collected during esophageal cancer surgery. L-CTCs were detected in 30% of patients (4/13). Interestingly, L-CTC achieved xenotransplantation rates of more than 60% (8/13) [87]. Furthermore, the invasive potential of L-CTC source cell lines was higher than that of ordinary cell lines [1, 85, 86] (Fig. 6.2).

6.9 CTC Cell Lines from Companies

Jacob et al. [88] used pancreatic cancer CTC cell lines (CM61, CF49, and HM59) obtained from Celprogen. Cell lines were generated from the following patients: a Caucasian male aged 61 years (CM61), a Caucasian female aged 49 years (CF49), and a Hispanic male aged 59 years (HM59). CTCs were isolated from the peripheral blood of PDAC patients with liver metastasis. However, other information was not provided. As indicated in the section on CTC cell lines, a pancreatic cancer CTC cell line has not yet been established. Therefore, it has not been confirmed whether these cell lines are authentic CTC cell lines, and their reliability has yet to be guaranteed.

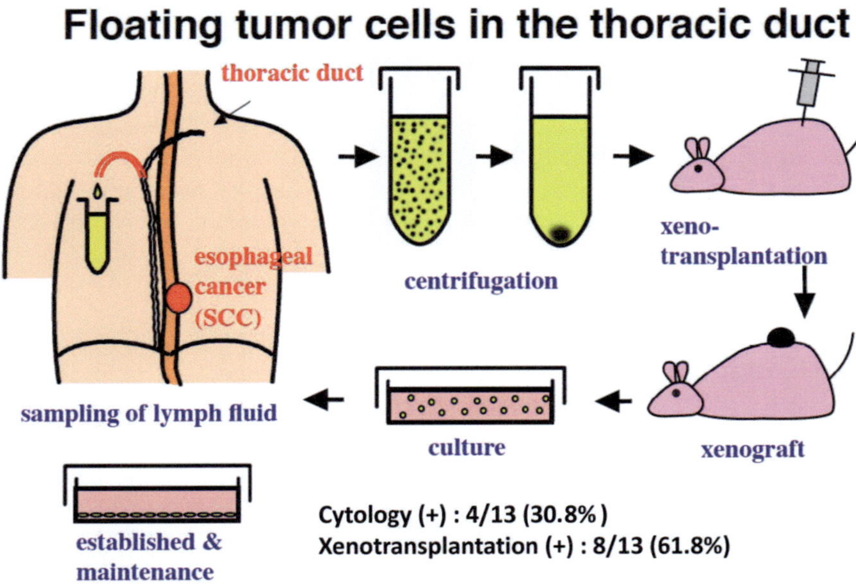

Fig. 6.2 Establishment of cancer cell lines derived from the thoracic duct lymph of esophageal cancer. These cell lines exhibited a high migratory ability in vitro and high lymph node metastatic potential in vivo [1]

Signorelli et al. [89] used the same CTC cell lines. They identified several novel surface biomarkers on CTCs expressed under different culture conditions. ALCAM (CD166) was validated and identified as a novel alternate surface marker on EpCAM-low CTCs in patients with metastatic pancreatic cancer.

CM61 CTCs were grown under low attachment or adherent conditions and differentiated into two subtypes of cells in cultures that required single cell sorting to isolate the cell types. These CTC lines exhibit the characteristics of leukocyte-like cells, which is consistent with the previous findings on CTCs with a hybrid double-positive phenotype.

Only cells that were cultured under low attachment conditions were used in co-cultures to mimic actual conditions in the blood. When CTCs were cultured under adherent and low attachment conditions, the expression of most proteins remained unchanged, even though differences were observed in some surface proteins. The most pronounced difference was observed when a CTC cell line was cultured with naïve neutrophils.

6.10 Technical Movie and Protocol

A previous study provided a technical movie and protocol for a CTC cell line [90].

6.11 Conclusion

In contrast to studies on circulating nucleic acids, exosomes, and other blood biomarkers, analyses of CTCs define their cellular genetics, epigenetics, transcriptomics, and protein profiles. However, CTC cultures are still challenging. The main limitation is the small number of CTCs in the bloodstream, which makes their isolation and enrichment paramount prerequisites, with different approaches available. The isolation approach is not standardized, and different methodologies for the characterization of CTCs may result in discrepancies. Furthermore, there is currently no gold standard for CTC cultures. The best approach may be achieved by combining several methods. Despite the difficulties associated with establishing cell lines from CTCs, short-term cultures of CTCs will provide valuable information.

Acknowledgments We thank Mr. Kiyoshi Okano, Mr. Tetsuo Sudo, and Mr. Akira Myomoto for their useful comments. We also thank Professor Shin Yonehara for his advice on growth supplements.

Funding The authors declare financial support from Toray Industries Inc.

Conflicts of Interest The authors declare no conflicts of interest.

References

1. Shimada Y, Sato F. Molecular factors related to metastasis of esophageal squamous cell carcinoma. Esophagus. 2007;4:7–18.
2. Shimada Y, Sudo T, Akamatsu S, et al. Cell lines of circulating tumor cells: what is known and what needs to be resolved. J Pers Med. 2022;12(5):666.
3. Grillet F, Bayet E, Villeronce O, et al. Circulating tumour cells from patients with colorectal cancer have cancer stem cell hallmarks in ex vivo culture. Gut. 2017;66(10):1802–10.
4. Soler A, Cayrefourcq L, Mazard T, et al. Autologous cell lines from circulating colon cancer cells captured from sequential liquid biopsies as model to study therapy-driven tumor changes. Sci Rep. 2018;8(1):15931.
5. Stickler S, Rath B, Hochmair M, et al. Changes of protein expression during tumorosphere formation of small cell lung cancer circulating tumor cells. Oncol Res. 2023;31(1):13–22.
6. Rath B, Plangger A, Krenbek D, et al. Rovalpituzumab tesirine resistance: analysis of a corresponding small cell lung cancer and circulating tumor cell line pair. Anti-Cancer Drugs. 2022;33(3):300–7.
7. Belthier G, Homayed Z, Grillet F, et al. CD44v6 defines a new population of circulating tumor cells not expressing EpCAM. Cancers (Basel). 2021;13(19):4966.
8. Cayrefourcq L, Mazard T, Joosse S, et al. Establishment and characterization of a cell line from human circulating colon cancer cells. Cancer Res. 2015;75(5):892–901.
9. Okumura T, Shimada Y, Omura T, et al. MicroRNA profiles to predict postoperative prognosis in patients with small cell carcinoma of the esophagus. Anticancer Res. 2015;35(2):719–27.
10. May AN, Crawford BD, Nedelcu AM. In vitro model-systems to understand the biology and clinical significance of circulating tumor cell clusters. Front Oncol. 2018;8:63.

11. Feng H, Ou BC, Zhao JK, et al. Homogeneous pancreatic cancer spheroids mimic growth pattern of circulating tumor cell clusters and macrometastases: displaying heterogeneity and crater-like structure on inner layer. J Cancer Res Clin Oncol. 2017;143(9):1771–86.
12. Park JY, Jeong AL, Joo HJ, et al. Development of suspension cell culture model to mimic circulating tumor cells. Oncotarget. 2017;9(1):622–40.
13. Huh HD, Sub Y, Oh J, et al. Reprogramming anchorage dependency by adherent-to-suspension transition promotes metastatic dissemination. Mol Cancer. 2023;22(1):63.
14. Arnoletti JP, Fanaian N, Reza J, et al. Pancreatic and bile duct cancer circulating tumor cells (CTC) form immune-resistant multi-cell type clusters in the portal venous circulation. Cancer Biol Ther. 2018;19(10):887–97.
15. Balakrishnan A, Koppaka D, Anand A, et al. Circulating tumor cell cluster phenotype allows monitoring response to treatment and predicts survival. Sci Rep. 2019;9(1):7933.
16. De T, Goyal S, Balachander G, et al. A novel ex vivo system using 3D polymer scaffold to culture circulating tumor cells from breast cancer patients exhibits dynamic E-M phenotypes. J Clin Med. 2019;8(9):1473.
17. Lee HL, Chiou JF, Wang PY, et al. Ex vivo expansion and drug sensitivity profiling of circulating tumor cells from patients with small cell lung cancer. Cancers (Basel). 2020;12(11):3394.
18. Ju L, Yang J, Zhai C, et al. Chemotherapy and chemosensitivity predicted by CTC cultured in vitro of SCLC patients. Front Oncol. 2021;11:683318.
19. Lin KC, Ting LL, Chang CL, et al. Ex vivo expanded circulating tumor cells for clinical anti-cancer drug prediction in patients with head and neck cancer. Cancers (Basel). 2021;13(23):6076.
20. Hu CL, Zhang YJ, Zhang XF, et al. 3D culture of circulating tumor cells for evaluating early recurrence and metastasis in patients with hepatocellular carcinoma. Onco Targets Ther. 2021;14:2673–88.
21. Wu YH, Hung YP, Chiu NC, et al. Correlation between drug sensitivity profiles of circulating tumour cell-derived organoids and clinical treatment response in patients with pancreatic ductal adenocarcinoma. Eur J Cancer. 2022;166:208–18.
22. Arnoletti JP, Reza J, Rosales A, et al. Pancreatic ductal adenocarcinoma (PDAC) circulating tumor cells influence myeloid cell differentiation to support their survival and immunoresistance in portal vein circulation. PLoS One. 2022;17(3):e0265725.
23. Pizon M, Schott D, Pachmann U, et al. Chick chorioallantoic membrane (CAM) assays as a model of patient-derived xenografts from circulating cancer stem cells (cCSCs) in breast cancer patients. Cancers (Basel). 2022;14(6):1476.
24. Morita K, Sawabata N, Tatsumi S, et al. Verification of a new filter for isolation of circulating tumor cells by only blood filtration. Anticancer Res. 2022;42(9):4305–10.
25. Kwizera EA, Ou W, Lee S, et al. Greatly enhanced CTC culture enabled by capturing CTC heterogeneity using a PEGylated PDMS-titanium-gold electromicrofluidic device with glutathione-controlled gentle cell release. ACS Nano. 2022;16(7):11374–91.
26. Hsieh CH, Yeh CT, Huang YH, et al. Circulating tumor cells derived from advanced hepatocellular carcinoma rapidly develop resistance to cytotoxic chemotherapy. Anticancer Res. 2022;42(5):2479–86.
27. Vardas V, Ju JA, Christopoulou A, et al. Functional analysis of viable circulating tumor cells from triple-negative breast cancer patients using TetherChip technology. Cells. 2023;12(15):1940.
28. Balakrishnan A, Thottian AGF, Govind Babu K, et al. Drug susceptibility testing of circulating lung cancer cells for personalized treatment. Med Oncol. 2022;40(1):1.
29. Nomura M, Yokoyama Y, Yoshimura D, et al. Simple detection and culture of circulating tumor cells from colorectal cancer patients using poly(2-methoxyethyl acrylate)-coated plates. Int J Mol Sci. 2023;24(4):3949.
30. Xiao J, Sharma U, Arab A, et al. Propagated circulating tumor cells uncover the potential role of NFκB, EMT, and TGFβ signaling pathways and COP1 in metastasis. Cancers (Basel). 2023;15(6):1831.

31. Hong B, Zhang X, Du X, et al. Exploring the potential driver gene mutations that promote renal cancer cell metastasis and implantation based on circulating tumor cells culture. Diagnostics (Basel). 2023;13(11):1855.
32. Mohamed BM, Ward MP, Bates M, et al. Ex vivo expansion of circulating tumour cells (CTCs). Sci Rep. 2023;13(1):3704.
33. Furukawa A, Mori T, Shimomura O, et al. Culture of circulating tumor cells using a microfilter device. Oncol Rep. 2023;49(5):101.
34. Wang J, Meng X, Yu M, et al. A novel microfluidic system for enrichment of functional circulating tumor cells in cancer patient blood samples by combining cell size and invasiveness. Biosens Bioelectron. 2023;227:115159.
35. Espejo-Cruz ML, González-Rubio S, Espejo JJ, et al. Enumeration and characterization of circulating tumor cells in patients with hepatocellular carcinoma undergoing transarterial chemoembolization. Int J Mol Sci. 2023;24(3):2558.
36. Che TF, Chiu CH, Wu YC, et al. Proliferative ability of circulating tumor cells is a prognostic factor in early-stage lung adenocarcinoma. Lung Cancer. 2023;178:198–205.
37. Allard WJ, Matera J, Miller MC, et al. Tumor cells circulate in the peripheral blood of all major carcinomas but not in healthy subjects or patients with nonmalignant diseases. Clin Cancer Res. 2004;10(20):6897–904.
38. Yeo D, Bastian A, Strauss H, et al. Exploring the clinical utility of pancreatic cancer circulating tumor cells. Int J Mol Sci. 2022;23(3):1671.
39. Malkawi W, Lutfi A, Afghan MK, et al. Circulating tumour cell enumeration, biomarker analyses, and kinetics in patients with colorectal cancer and other GI malignancies. Front Oncol. 2023;13:1305181.
40. Bissolati M, Sandri MT, Burtulo G, et al. Portal vein-circulating tumor cells predict liver metastases in patients with resectable pancreatic cancer. Tumour Biol. 2015;36(2):991–6.
41. Catenacci DV, Chapman CG, Xu P, et al. Acquisition of portal venous circulating tumor cells from patients with pancreaticobiliary cancers by endoscopic ultrasound. Gastroenterology. 2015;149(7):1794–1803.e4.
42. White MG, Lee A, Vicente D, et al. Measurement of portal vein blood circulating tumor cells is safe and may correlate with outcomes in resected pancreatic ductal adenocarcinoma. Ann Surg Oncol. 2021;28(8):4615–22.
43. Stoecklein NH, Fluegen G, Guglielmi R, et al. Ultra-sensitive CTC-based liquid biopsy for pancreatic cancer enabled by large blood volume analysis. Mol Cancer. 2023;22(1):181.
44. Gardner KP, Aldakkak M, Tang CM, et al. Circulating stromal cells in resectable pancreatic cancer correlates to pathological stage and predicts for poor clinical outcomes. NPJ Precis Oncol. 2021;5(1):25.
45. Cheng SB, Chen MM, Wang YK, et al. A three-dimensional conductive scaffold microchip for effective capture and recovery of circulating tumor cells with high purity. Anal Chem. 2021;93(18):7102–9.
46. Mendelaar PAJ, Kraan J, Van M, et al. Defining the dimensions of circulating tumor cells in a large series of breast, prostate, colon, and bladder cancer patients. Mol Oncol. 2021;15(1):116–25.
47. Yu M, Bardia A, Aceto N, et al. Cancer therapy. Ex vivo culture of circulating breast tumor cells for individualized testing of drug susceptibility. Science. 2014;345(6193):216–20.
48. Hamilton G, Rath B, Stickler S. Significance of circulating tumor cells in lung cancer: a narrative review. Transl Lung Cancer Res. 2023;12(4):877–94.
49. Kolostova K, Cegan M, Bobek V. Circulating tumour cells in patients with urothelial tumours: enrichment and in vitro culture. Can Urol Assoc J. 2014;8(9–10):E715–20.
50. Kolostova K, Broul M, Schraml J, et al. Circulating tumor cells in localized prostate cancer: isolation, cultivation in vitro and relationship to T-stage and Gleason score. Anticancer Res. 2014;34(7):3641–6.

51. Kitz J, Goodale D, Postenka C, et al. EMT-independent detection of circulating tumor cells in human blood samples and pre-clinical mouse models of metastasis. Clin Exp Metastasis. 2021;38(1):97–108. https://doi.org/10.1007/s10585-020-10070-y. Epub 2021 Jan 7.
52. Maeshiro M, Shinriki S, Liu R, et al. Colonization of distant organs by tumor cells generating circulating homotypic clusters adaptive to fluid shear stress. Sci Rep. 2021;11(1):6150.
53. Agerbæk MØ, Bang-Christensen SR, Yang MH, et al. The VAR2CSA malaria protein efficiently retrieves circulating tumor cells in an EpCAM-independent manner. Nat Commun. 2018;9(1):3279.
54. Gao Y, Wu M, Luan Q, et al. Acoustic bubble for spheroid trapping, rotation, and culture: a tumor-on-a-chip platform (ABSTRACT platform). Lab Chip. 2022;22(4):805–13.
55. Xiang Y, Zhang H, Lu H, et al. Bioorthogonal microbubbles with antifouling nanofilm for instant and suspended enrichment of circulating tumor cells. ACS Nano. 2023;17(10):9633–46.
56. Geng W, Liu Y, Yu N, et al. An ultra-compact acoustofluidic device based on the narrow-path travelling surface acoustic wave (np-TSAW) for label-free isolation of living circulating tumor cells. Anal Chim Acta. 2023;1255:341138.
57. Kantara C, O'Connell MR, Luthra G, et al. Methods for detecting circulating cancer stem cells (CCSCs) as a novel approach for diagnosis of colon cancer relapse/metastasis. Lab Investig. 2015;95(1):100–12.
58. Khoo BL, Lee SC, Kumar P, et al. Short-term expansion of breast circulating cancer cells predicts response to anti-cancer therapy. Oncotarget. 2015;6(17):15578–93.
59. Gao D, Vela I, Sboner A, et al. Organoid cultures derived from patients with advanced prostate cancer. Cell. 2014;159(1):176–87.
60. Hamilton G, Burghuber O, Zeillinger R. Circulating tumor cells in small cell lung cancer: ex vivo expansion. Lung. 2015;193(3):451–2.
61. Zhang L, Ridgway LD, Wetzel MD, et al. The identification and characterization of breast cancer CTCs competent for brain metastasis. Sci Transl Med. 2013;5(180):180ra48.
62. Khoo BL, Grenci G, Lim YB, et al. Expansion of patient-derived circulating tumor cells from liquid biopsies using a CTC microfluidic culture device. Nat Protoc. 2018;13(1):34–58.
63. Xiao J, McGill JR, Stanton K, et al. Efficient propagation of circulating tumor cells: a first step for probing tumor metastasis. Cancers (Basel). 2020;12(10):2784.
64. Zhang Z, Shiratsuchi H, Lin J, et al. Expansion of CTCs from early stage lung cancer patients using a microfluidic co-culture model. Oncotarget. 2014;5(23):12383–97.
65. Rousset X, Maillet D, Grolleau E, et al. Embryonated chicken tumor xenografts derived from circulating tumor cells as a relevant model to study metastatic dissemination: a proof of concept. Cancers (Basel). 2022;14(17):4085.
66. Arora S, D'Souza A, Aland G, et al. Antibody mediated cotton-archetypal substrate for enumeration of circulating tumor cells and chemotherapy outcome in 3D tumors. Lab Chip. 2022;22(8):1519–30.
67. Carmona-Ule N, González-Conde M, Abuín C, et al. Short-term ex vivo culture of CTCs from advance breast cancer patients: clinical implications. Cancers (Basel). 2021;13(11):2668.
68. Andree KC, Mentink A, Zeune LL, et al. Toward a real liquid biopsy in metastatic breast and prostate cancer: diagnostic LeukApheresis increases CTC yields in a European prospective multicenter study (CTCTrap). Int J Cancer. 2018;143(10):2584–91.
69. Fehm TN, Meier-Stiegen F, Driemel C, et al. Diagnostic leukapheresis for CTC analysis in breast cancer patients: CTC frequency, clinical experiences and recommendations for standardized reporting. Cytometry A. 2018;93(12):1213–9.
70. Mout L, van Dessel LF, Kraan J, et al. Generating human prostate cancer organoids from leukapheresis enriched circulating tumour cells. Eur J Cancer. 2021;150:179–89.
71. Tamminga M, Andree KC, Hiltermann TJN, et al. Detection of circulating tumor cells in the diagnostic leukapheresis product of non-small-cell lung cancer patients comparing CellSearch(®) and ISET. Cancers (Basel). 2020;12(4):896.
72. Lambros MB, Seed G, Sumanasuriya S, et al. Single-cell analyses of prostate cancer liquid biopsies acquired by apheresis. Clin Cancer Res. 2018;24(22):5635–44.

73. Franken A, Driemel C, Behrens B, et al. Label-free enrichment and molecular characterization of viable circulating tumor cells from diagnostic leukapheresis products. Clin Chem. 2019;65(4):549–58.
74. Magbanua MJ, Pugia M, Lee JS, et al. A novel strategy for detection and enumeration of circulating rare cell populations in metastatic cancer patients using automated microfluidic filtration and multiplex immunoassay. PLoS One. 2015;10(10):e0141166.
75. Ilie M, Hofman V, Long-Mira E, et al. "Sentinel" circulating tumor cells allow early diagnosis of lung cancer in patients with chronic obstructive pulmonary disease. PLoS One. 2014;9(10):e111597.
76. Pantel K, Denève E, Nocca D, et al. Circulating epithelial cells in patients with benign colon diseases. Clin Chem. 2012;58(5):936–40.
77. Terai M, Mu Z, Eschelman DJ, et al. Arterial blood, rather than venous blood, is a better source for circulating melanoma cells. EBioMedicine. 2015;2(11):1821–6.
78. Weissenstein U, Schumann A, Reif M, et al. Detection of circulating tumor cells in blood of metastatic breast cancer patients using a combination of cytokeratin and EpCAM antibodies. BMC Cancer. 2012;12:206.
79. Fang ZT, Zhang W, Wang GZ, et al. Circulating tumor cells in the central and peripheral venous compartment—assessing hematogenous dissemination after transarterial chemoembolization of hepatocellular carcinoma. Onco Targets Ther. 2014;7:1311–8.
80. Jiao LR, Apostolopoulos C, Jacob J, et al. Unique localization of circulating tumor cells in patients with hepatic metastases. J Clin Oncol. 2009;27(36):6160–5.
81. Rahbari NN, Bork U, Kircher A, et al. Compartmental differences of circulating tumor cells in colorectal cancer. Ann Surg Oncol. 2012;19(7):2195–202.
82. Denève E, Riethdorf S, Ramos J, et al. Capture of viable circulating tumor cells in the liver of colorectal cancer patients. Clin Chem. 2013;59(9):1384–92.
83. Diamantopoulou Z, Castro-Giner F, Schwab FD, et al. The metastatic spread of breast cancer accelerates during sleep. Nature. 2022;607(7917):156–62.
84. Micalizzi DS, Maheswaran S, Haber DA. A conduit to metastasis: circulating tumor cell biology. Genes Dev. 2017;31(18):1827–40.
85. Sato F, Shimada Y, Watanabe G, et al. Characterization of esophageal cancer cell lines derived from lymph fluid in thoracic duct. Can J Gastroenterol. 1998;12:106B.
86. Nakaji T, Kataoka TR, Watabe K, et al. A new member of the GTPase superfamily that is upregulated in highly metastatic cells. Cancer Lett. 1999;147(1–2):139–47.
87. Watanabe H, Tachimori Y, Katon H, et al. Tumor progression of cancer cells derived from lymph fluid of thoracic duct for esophageal cancer patients by orthotopic transplantation on nude rat. Biotherapy. 1996;10:1332–7.
88. Charles Jacob HK, Charles Richard JL, Signorelli R, et al. Modulation of early neutrophil granulation: the circulating tumor cell-extravesicular connection in pancreatic ductal adenocarcinoma. Cancers (Basel). 2021;13(11):2727.
89. Signorelli R, Giret TM, Umland O, et al. ALCAM: a novel surface marker on EpCAM(low) circulating tumor cells. Biomedicine. 2022;10(8):1983.
90. Belthier G, Homayed Z, Bouclier C, et al. Circulating tumor cell lines: an innovative tool for fundamental and translational research. J Vis Exp. 2021;(178).

Chapter 7
Patient-Derived Organoids: Basic and Translational Research on Uterine Cancer

Yoshiaki Maru, Naotake Tanaka, and Yoshitaka Hippo

Abstract Various carcinogenesis steps, including tumor development, progression, metastasis, and recurrence, contribute to cancer heterogeneity. Heterogeneity may lead to drug resistance in cancer treatment; therefore, it is crucial to establish cancer models that can retain the biological properties of human tumors, including heterogeneity, for elucidating the mechanisms underlying cancer pathogenesis, validating candidate drugs, and implementing precision medicine. Owing to its ability to retain the histological and molecular characteristics of the original tumor and its high success rate in primary cancer cell culture, organoid culture has garnered considerable attention. Consequently, patient-derived organoids have been generated from the clinical samples of various cancer types. Since we have pioneered the organoid-based research field, particularly in gynecologic cancers, herein, we provide an overview of the organoid culture and its applications in cancer research, focusing on the current status of organoid-based research on endometrial and cervical cancers. Additionally, the challenges and perspectives associated with organoid culture techniques in basic and translational cancer research have been discussed.

Keywords Organoid · Endometrial cancer · Cervical cancer · Heterogeneity

Y. Maru (✉) · Y. Hippo
Laboratory of Precision Tumor Model Systems, Chiba Cancer Center Research Institute, Chiba, Japan
e-mail: ymaru@chiba-cc.jp; yhippo@chiba-cc.jp

N. Tanaka
Department of Gynecology, Chiba Cancer Center, Chiba, Japan

N. Shinomiya, Y. Shimada (eds.), *Applications of Cell Culture*, Current Human Cell Research and Applications, https://doi.org/10.1007/978-981-95-0587-6_7

7.1 Introduction

7.1.1 Overview

Organoids are self-organizing three-dimensional (3D) cellular structures derived from differentiating, self-renewing stem and/or progenitor cells, namely, pluripotent stem cells (PSCs), such as embryonic stem cells and induced PSCs or somatic stem cells from adult tissues [1]. PSC-derived organoids typically contain mixed cell types and exhibit a complex morphology, making them preferable in organ development and regenerative medicine studies [2, 3]. In contrast, somatic stem cell-derived organoids, which contain only epithelial cells, are mostly used in cancer research. In 2009, the infinite proliferation of LGR5+-marked normal murine intestinal stem cells was demonstrated in vitro; the cells differentiated into all lineages of the small intestine [4]. Specifically, the in vivo intestinal homeostasis was reconstituted with the extracellular matrix (ECM) and defined stem cell niche factors without the aid of a stromal niche. Thereafter, this protocol has been optimized for various normal murine and human organs [5] and extensively applied in various research fields including regenerative medicine [6], bacterial infection [7], and cancer [8]. Technically, because cancer does not represent an organ, using the term "organoid" for cancer cells may not be appropriate; therefore, tumor-derived organoids are occasionally referred to as "tumoroids." Nevertheless, most researchers accept the use of "organoid" for cancer cells, considering the similarity of cancer stem cells to tissue stem cells. Accordingly, patient-derived cells propagated in ECM, such as Matrigel, have also been described as organoids herein. Notably, some researchers have referred to cell aggregates of patient-derived cells and cancer cell lines cultured in floating culture as "organoids," which may be better described as spheroids. However, even 3D, round-shaped structures under the ECM-embedded state could be referred to as "spheroid." Such nomenclature issues must be considered when examining the literature.

7.1.2 Procedure of Organoid Culture

The "dome culture" method is the original and most common organoid culture method, in which cells are mixed with ECM and solidified on the plate (Fig. 7.1; upper panel). Briefly, tissues that are physically minced into small pieces using scissors and a scalpel are enzymatically dissociated, typically using collagenase and/or dispase. Next, dissociated cells are resuspended in liquid Matrigel and plated in drops onto a pre-warmed culture plate. After solidifying, Matrigel drops are cultured in media supplemented with defined growth factors and signaling pathway inhibitors. Alternatively, we developed the Matrigel bilayer organoid culture (MBOC) protocol, which was originally developed to achieve high viral infection efficiency in organoids (Fig. 7.1; lower panel) and could also be used for primary

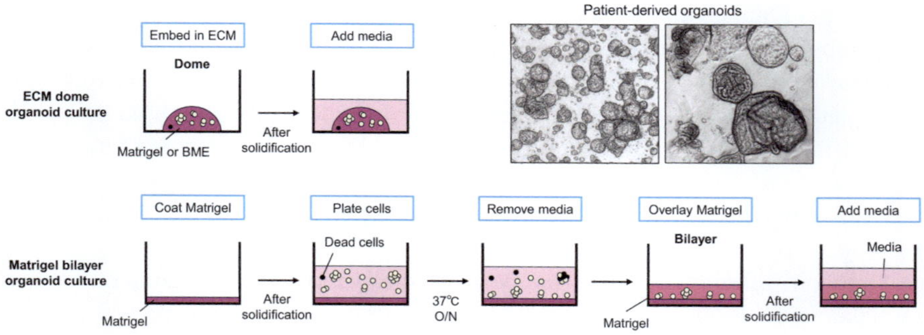

Fig. 7.1 Representative two methods of organoid culture. Organoid cultures propagate in both normal and cancer cells while retaining differentiation and heterogeneity. *Upper* panel: ECM dome organoid cultures. ECM, extracellular matrix; BME, basement membrane extract. *Lower* panel: Matrigel bilayer organoid culture

culture and subculture [9]. MBOC has been adopted as a standard organoid culture method for various epithelial cells in our research because it allows the efficient removal of dead or differentiated cells, which could digest Matrigel or exert adverse effects on viable cells. The culture process is the same as that in the conventional dome culture method until the step of enzymatic dissociation of tissues. The dissociated cells resuspended in organoid culture media are plated on a layer of solidified Matrigel and incubated overnight at 37 °C. Thereafter, floating or dead cells are discarded along with media, and the viable cells attached to the Matrigel are covered with Matrigel and overlaid with culture media. Using this approach, we have successfully established organoid-based ex vivo/in vivo hybrid carcinogenesis models for various organs [10, 11]. For subculture, the propagated organoids are enzymatically and physically dissociated into single cells or small-cell aggregates, which are then subjected to organoid culture using the aforementioned dome or bilayer culture methods. Many organoids can tolerate cryopreservation for subsequent thawing and reculturing. Additionally, we have prepared organoid-derived spheroids for drug sensitivity and high-throughput drug screening because Matrigel embedding-based in vitro assays are time-consuming and require complicated procedures. Specifically, organoids dissociated into single cells were resuspended in a 2% Matrigel/organoid culture medium and plated onto ultralow-attachment plates.

7.1.3 Application in Cancer Research

Accumulation of genetic aberrations in normal epithelial cells can drive carcinogenesis, which is further augmented by the microenvironment [12, 13]. During progression, metastasis, and recurrence, tumors can give rise to a heterogeneous cell population, thereby leading to drug resistance development [14]. Therefore, establishing patient-derived cancer models that accurately recapitulate the biological properties of human tumors is vital for elucidating the mechanisms underlying cancer pathogenesis, validating candidate drugs, and implementing precision medicine. Various approaches, including patient-derived xenografts, floating spheroid cultures, and organoid cultures, have been reported for developing such models [15]. Among them, organoid culture has garnered notable attention because of its advantages over other models; it can facilitate the propagation of normal and precancerous cells, enable a higher establishment rate for cancer cells, and retain tumor heterogeneity.

Organoid culture techniques can be applied to various clinical samples such as resected tissue, biopsy, ascites, and pleural effusions. Accordingly, patient-derived organoids (PDOs) have been established for various cancer types, including colon [16], pancreatic [17], gastric [18], breast [19], bladder [20], esophagus [21], and lung cancer [22]. Notably, the MBOC protocol exhibited good efficacy for the organoid culture of colon cancer tissues and cancerous ascites. In contrast, establishing PDOs from ovarian cancer tissues was challenging owing to their inherent enzymatic dissociation resistance. To circumvent this issue, the MBOC protocol was modified to achieve a more intensive dissociation of both cancer and stromal cells from ovarian tumor tissues [23]. Consequently, the modified MBOC protocol was successfully applicable for establishing PDOs from endometrial, cervical, and ovarian cancers [23, 24].

Propagated PDOs can retain the histological and genetic characteristics of their original tumors, making them feasible for in vitro drug sensitivity assays to potentially mirror the clinical responses of patients [25]. Therefore, PDOs may be crucial clinical response-predicting resources for the implementation of precision medicine [26]. Many clinical trials are underway to investigate if PDO can predict patient responses to treatment [27]. Furthermore, PDOs have been used for the high-throughput screening of compounds. For instance, a compound library consisting of clinically used 339 drugs was screened for repurposing based on the ability of compounds to suppress PDOs of biliary tract carcinoma to identify candidate therapeutic drugs with minimal toxicity to normal biliary epithelial cells [28]. Furthermore, screening of 37 anticancer drugs for PDO of gastric cancer revealed potential targets for personalized therapy [29]. Additionally, multiple genetic alterations and/or transcription factors have been reconstituted in PDOs using CRISPR-Cas9 technology [30]. These models may be useful in elucidating the molecular mechanisms underlying carcinogenesis. Most studies adopt a single PDO as an avatar for the patient; however, tumors frequently display spatial diversity, potentially questioning the validity of single-PDO-derived results. Reportedly, PDOs originating from different regions of colon cancer per patient have shown heterogeneous drug responses [31]. Moreover, multiple PDOs from the primary and disseminated foci of the same patient with ovarian cancer have been reported to present a close association between tumor heterogeneity and drug sensitivity [32]. Therefore, further studies on the representativeness of a single PDO for the original tumor are warranted in large-scale cohorts of various cancer types. Additionally, PDOs have been used for various analyses, including xenograft development and single-cell RNA sequencing [33].

7.2 Organoid-Based Research on Endometrial Cancer

7.2.1 *Background*

Endometrial cancer is the most common cancer of the female reproductive tract worldwide [34], with a 5-year overall survival rate of 74%–90% for patients with stage I and II endometrial cancer and 20–26% for patients with stage IV endometrial cancer [35] based on FIGO staging. It is mainly divided into two distinct subtypes (Type I and II) based on its clinical and pathological characteristics. Type I endometrial cancer, comprising 80% of all endometrial cancers, is mainly associated with estrogen dependency, endometrial hyperplasia, and a favorable prognosis. The predominant histological subtype is endometrioid carcinoma, which mostly harbors mutations in PTEN, followed by those in PIK3CA, KRAS, ARID1A, and CTNNB1. In contrast, Type II endometrial cancer comprises various minor subtypes, including serous carcinoma, clear cell carcinoma, and carcinosarcoma, and it is associated with estrogen independence, endometrial atrophy, and poor prognoses. Mutations in TP53, PIK3CA, and PPP2R1A, and ERBB2 amplification are among the most frequent genetic alterations in Type II subtype [36]. Based on the comprehensive genomic profiling of endometrial cancer, four molecular subtypes have recently been proposed: DNA polymerase epsilon ultramutated, microsatellite instability hypermutated, copy number low, and copy number high [37]. Although recent molecular targeted therapies, including antiangiogenic agents and immune checkpoint inhibitors, have exhibited efficacy in a subset of patients with advanced or recurrent disease [38], treatment options for endometrial cancer are limited, underscoring the need for the development of new therapeutic strategies. To address this issue, patient-derived cancer models are extensively established. Representative PDO-based studies on endometrial cancer are summarized in Table 7.1.

7.2.2 *Tumor-Derived Organoids*

Studies on PDOs of endometrial cancer have mainly focused on endometrioid carcinoma, which is the most common histological type [39]. We have also established endometrioid carcinoma-PDOs using a modified MBOC protocol [23]. In 2019, 14 PDOs were established for endometrial cancer [40], which retained the phenotypic and genetic heterogeneities of their original tumors. Higher expression of WNT target genes was observed in low-grade endometrial carcinoma than that in high-grade endometrial carcinoma. In contrast, PDOs of high-grade endometrial carcinoma presented downregulated ESR1 and FOXA2 and upregulated epithelial–mesenchymal transition-associated genes. Additionally, PDO with mutations in PTEN, PIK3CA, and AKT1 exhibited highest sensitivity to everolimus, a mTOR inhibitor, suggesting a strong dependence on the PI3K/AKT pathway. A drug sensitivity assay was also performed using eight PDOs generated from

Table 7.1 Representative organoid-based research on endometrial cancer

Sampling site	Histology	Examination	ECM	Composition of media[a]	References
Carcinoma	EMC	Characterization, drug sensitivity	BME	B27, *N*-acetyl-L-cysteine, Nicotinamide, A83-01, SB202190, Y-27632, 17-β estradiol	Girda et al. [39]
Carcinoma	EMC	Characterization, growth assay, ODX	Matrigel	EGF, Y-27632, Noggin, R-spondin1, Jagged-1	Maru et al. [23]
Carcinoma	EMC, SC, CCC	Characterization, drug sensitivity, ODX, ion channel functionality	Matrigel	R-spondin1 CM, Noggin, B27, N2, Nicotinamide, A83-01, *N*-acetyl-L-cysteine, EGF, IGF-1, HGF, SB202190, 17-β estradiol, Y-27632[b]	Boretto et al. [40]
Carcinoma	EMC, SC, mixed SC/EMC	Drug sensitivity	Matrigel	B27, *N*-acetyl-L-cysteine, R-spondin1, Noggin, Nicotinamide, EGF, FGF10, Forskolin, A83-01, Hydrocortisone, Heregulinβ-1, 17-β-estradiol, Y-27632	Bi et al. [41]
Carcinoma	EMC, SC, CCC, CS	Characterization, drug sensitivity, ODX	Matrigel	B27, EGF, Noggin, Y-27632, 17-β estradiol, SB202190, Nicotinamide, A83-01, R-spondin1, *N*-acetyl-L-cysteine	Berg et al. [42]
Carcinoma	No description	Growth assay	Matrigel	WRN-CM, B27, N2, Insulin-transferrin-selenium, Primocin, EGF, FGF10, *N*-acetyl-L-cysteine, Nicotinamide, Estrogen, A83-01	Jamaluddin et al. [43]
Carcinoma	CS	Characterization, drug sensitivity, drug screening	Matrigel	EGF, Y-27632, Noggin, R-spondin1, Jagged-1	Maru et al. [44]
Benign	Hyperplasia	Characterization	Matrigel	R-spondin1 CM, Noggin, B27, N2, Insulin-transferrin-selenium, Nicotinamide, A83-01, *N*-acetyl-L-cysteine, EGF, bFGF, FGF-10, SB202190, 17-β estradiol[c], Y-27632[b]	Boretto et al. [40]

(continued)

Table 7.1 (continued)

Sampling site	Histology	Examination	ECM	Composition of media[a]	References
Normal		Characterization, hormone response	Matrigel	B27, N2, Insulin-transferrin-selenium, Nicotinamide, EGF, FGF10, Noggin, A83-01, *N*-acetyl-L-cysteine, SB202190, 17β-estradiol, Wnt3A CM, R-spondin1 CM	Boretto et al. [45]
Normal		Characterization, hormone response	Matrigel	A83-01, FGF10, HGF, EGF, Noggin, R-spondin1, Nicotinamide	Turco et al. [46]

EMC, endometrioid carcinoma; SC, serous carcinoma; CCC, clear cell carcinoma; CS, carcinosarcoma; ODX, organoid-derived xenograft; BME, basement membrane extract; EGF, epidermal growth factor; IGF, insulin-like growth factor; HGF, fibroblast growth factor; HGF, hepatocyte growth factor; CM, conditioned medium; WRN, Wnt3A-R-spondin3-Nogin

[a] Excluding basal media composition such as Glutamax, HEPES, and antibiotics

[b] Only for organoid formation or dissociation at passaging

[c] Only for organoid expansion

endometrial cancers, including endometrioid, serous, and mixed carcinoma [41]. These PDOs exhibited different sensitivities to standard cytotoxic and molecular targeted agents, with one PDO from a patient with serous carcinoma who had been treated with neoadjuvant trastuzumab before surgery successfully predicting post-operative chemotherapy and trastuzumab resistance. Furthermore, the PDO drug sensitivity assay identified alternative treatment options presently used in second-line settings. To explore the organoid phenotype associated with specific histological subtypes, multiplex immunohistochemical staining for PDOs has been performed using imaging mass cytometry [42]. The tSNE plots of single-cell segmentation masks of PDOs have identified two main groups corresponding to different histological subtypes. One group comprised endometrioid carcinoma grade 1–2, in which each PDO was independently distinguished, whereas the other group comprised endometrioid carcinoma grade 3 and non-endometrioid subtypes, in which each PDO was partially overlapped. These PDOs were orthotopically implanted into NSG mice, and 8 out of 10 PDO lines formed uterine tumors, yielding an engraftment rate of 80%. In general, tumor development is slower in lower-grade endometrioid tumors and more rapid in higher-grade endometrioid and non-endometrioid tumors.

Multiple PDOs were established in four cases from tissue samples of four different lesions from the same tumor [43]. Notably, morphological differences among the PDOs were observed irrespective of the size of the native tumor. The growth of these PDOs varied significantly, suggesting intratumoral heterogeneity. Recently, we established and characterized PDOs from multiple sources in a patient with advanced uterine carcinosarcoma, including resected tumor tissue, peritoneal lavage fluid, and intrauterine brushing of the tumor [44]. The three PDOs varied in many aspects, such as proliferation rate, gene amplification of KRAS and ERBB2, protein expression profiles, and sensitivity to standard cytotoxic agents and HER2 inhibitor. Despite the observed heterogeneity, drug screening identified four candidate reagents that were commonly effective for all PDOs. These studies suggest that multiple PDOs can help to reproduce the spatial diversity of a tumor and serve as a valuable resource in various endometrial cancer studies.

7.2.3 Nontumor-Derived Organoids

Estrogen and progesterone regulate the cyclical growth, differentiation, and shedding of normal endometrium; however, the underlying molecular and cellular mechanisms were poorly understood, primarily because of the lack of appropriate models. In 2017, human and mouse endometrial organoids were established [45, 46]. Compared with mouse endometrial organoids, WNT3A is not required for the stable expansion of human endometrial organoids, suggesting active endogenous WNT signaling pathways in humans. Consistent with this finding, certain WNT ligands have been expressed in human endometrial organoids. R-spondin1 is required for efficient and long-term expansion of human endometrial organoids, consistent with the high expression of LGR4 and LGR5 receptors for R-spondin1 [45]. Human endometrial organoids present a rounded shape, with a central lumen surrounded by an epithelial monolayer. Consistent with this finding, endometrial organoids express glandular markers, such as MUC1, FOXA2, and CDH1. Their gene expression profile exhibits more similarity to that of the glands than to that of the stroma, strongly suggesting its glandular epithelial nature [46]. Additionally, these studies describe the effects of successive estrogen (17β-estradiol, E2) and progesterone exposure in human endometrial organoids. E2 treatment has been reported to result in a higher proportion of Ki67-positive cells, with subsequent progesterone treatment reducing the number of proliferating cells. E2 treatment of endometrial organoids induces changes in morphology and gene expression, which mimics the proliferative phase of the endometrium, whereas progesterone administration induces secretory-phase endometrium-like changes [45]. Furthermore, endometrial organoids present a decidual-like phenotype characteristic of early pregnancy upon exposure to pregnancy signals such as placental hormones, prolactin, cyclic adenosine monophosphate, E2, and progesterone [46].

Endometrial hyperplasia, of which atypical endometrial hyperplasia often precedes cancer, is characterized by the proliferation of glands with an increased gland-to-stroma ratio compared to the normal proliferative endometrium; naturally, derived organoids exhibit high proliferative capacity. The columnar cells are lined in multiple layers on the luminal side of the organoids originating from the hyperplastic endometrium [40].

7.3 Organoid-Based Research on Cervical Cancer

7.3.1 Background

Cervical cancer is the third most common type of malignant tumor of the female genital system [34], with a persistent human papillomavirus (HPV) infection being the primary cause. Although incidences of cervical cancer are currently decreasing because of intensive cervical cancer screening and anti-HPV vaccinations, HPV-associated cervical cancer remains a challenge in many developing countries [47]. The most common histological subtype of cervical cancer is squamous cell carcinoma (SCC), accounting for 70% of all cases [48], while the remaining cases are predominantly of adenocarcinoma (ADC), the incidence of which is increasing worldwide and displays more aggressive features than SCC. ADC can be classified into several histological types. Among them, HPV-associated usual-type ADC is the most prevalent subtype, whereas HPV-independent gastric-type ADC is a rare subtype of ADC with an unfavorable prognosis [49]. HPV-independent cervical cancer presents different genomic profiles and a poorer prognosis compared with those of HPV-associated cervical cancer [50, 51]. Additionally, neuroendocrine carcinoma is a rare, aggressive subtype that accounts for approximately 2% of all cervical cancers. Based on its neuroendocrine features, chemotherapy for small cell lung cancer has been adopted. Therefore, the development of new therapeutic strategies for refractory cervical cancer is required. To address this issue, it is important to establish patient-derived cancer models. Representative PDO-based studies on cervical cancer are summarized in Table 7.2.

7.3.2 Tumor-Derived Organoids

Studies on organoid cultures of cervical cancer are limited. To the best of our knowledge, a PDO for clear cell carcinoma, a rare histological type of cervical cancer, was first established in 2019 [24]. In this case, MET gene amplification and overexpression of MET protein were detected in biopsy-derived PDO and primary tumor tissue from the same patient. MET-overexpressing PDO exhibited more sensitivity to MET inhibitors than PDO without MET amplification, suggesting the MET gene as a therapeutic target for recurring cancer. Using this PDO, the anticancer effects of a compound targeting a common mitochondrial DNA cancer risk variant were successfully evaluated [52]. In clinical practice, brush materials from the uterine cervix are used to detect precancerous and cancer cells. Twelve PDOs have been established from the brush materials of 11 patients with cervical cancer [53], which retained the genetic aberrations and causative HPV genomes of the original tumor. Further, they exhibit differential responses to common chemotherapeutic agents and achieve varying degrees of xenograft development in mice. These findings suggest the validity of PDOs as patient-derived cancer models, even if they originate from

Table 7.2 Representative organoid-based research on cervical cancer

Sampling site	Histology	Examination	ECM	Composition of media[a]	References
Carcinoma	CCC	Characterization, drug sensitivity	Matrigel	EGF, Y-27632, Noggin, R-spondin1, Jagged-1	Maru et al. [24]
Carcinoma	SCC, ADC	Characterization, drug sensitivity, ODX	BME	Noggin CM, Nicotinamide, SB202190, B27 supplement, Y-27632, FGF7, *N*-acetyl-L-cysteine, A83-01, Forskolin, FGF10, R-spondin1 CM	Lõhmussaar et al. [53]
Carcinoma	SCC, ADC, LCNEC, ADC (villoglandular)	Characterization, drug sensitivity, drug screening, radiosensitivity	Matrigel	Noggin, R-spondin-1 CM, B27 supplement, N2 supplement, EGF, HGF, Nicotinamide, *N*-acetyl-L-cysteine, Y-27632, A83-01, FGF10	Seol et al. [54]
Carcinoma	SCC, ADC, AdSq	Characterization, drug sensitivity, radiosensitivity, co-culture of PDOs and T cells, ODX	Matrigel	**SCC:** B27, N2, *N*-acetyl-L-cysteine, Nicotinamide, Noggin, A83-01, EGF, FGF10, Forskolin, Hydrocortisone **ADC:** B27, *N*-acetyl-L-cysteine, Nicotinamide, A83-01, R-spondin 1, Noggin, CHIR99021, EGF, FGF2, FGF7, FGF10, Prostaglandin E2	Huang et al. [55]

(continued)

Table 7.2 (continued)

Sampling site	Histology	Examination	ECM	Composition of media[a]	References
Carcinoma, Precancerous lesion	SCC, HSIL	Characterization, drug sensitivity, ODX, co-culture of PDOs and T cells	BME	**HSIL:** Noggin, FGF7, EGF, SB202190, Nicotinamide, *N*-acetyl-L-cysteine, Forskolin, B27, Y-27632, A83-01 **SCC:** Noggin, FGF7, SB202190, Nicotinamide, *N*-acetyl-L-cysteine, Forskolin, B27, Y-27632, A83-01	Hu et al. [56]
Carcinoma	HPV-asso CC, HPV-ind CC	Drug sensitivity	BME	No description	Wang et al. [57]
Normal	SCJ	Characterization	Matrigel	EGF, Y-27632, Noggin, R-spondin1, Jagged-1	Maru et al. [58]
Normal	Ectocervix, Endocervix	Characterization, viral infection	BME	**Ecto:** Noggin CM, Nicotinamide, SB202190, B27, Y-27632, FGF7, *N*-acetyl-L-cystein, A83-01, Forskolin, FGF10, R-spondin1 CM **Endo:** Noggin CM, Nicotinamide, SB202190, B27, Y-27632, FGF7, *N*-acetyl-L-cystein, A83-01, Forskolin, FGF10, R-spondin1 CM, EGF, 17β-estradiol, WNT surrogates, CHIR-99021	Lõhmussaar et al. [53]

(continued)

Table 7.2 (continued)

Sampling site	Histology	Examination	ECM	Composition of media[a]	References
Normal	Ectocervix, Endocervix	Characterization	Matrigel	**Ecto:** B27, N2, Hydrocortisone, EGF, Noggin, FGF10, *N*-acetyl-L-cysteine, Nicotinamide, SB431542, Y-27632, Forskolin **Endo:** B27, N2, EGF, Noggin, FGF10, *N*-acetyl-L-cysteine, Nicotinamide, SB431542, Y-27632, WNT3A CM, R-spondin CM	Chumduri et al. [60]
Nromal	Ectocervix	Characterization, viral infection	Matrigel	B27, N2, EGF, Hydrocortisone, Noggin, FGF10, *N*-acetyl-L-cysteine, SB431542, Y-27632, Nicotinamide, Forskolin	Koster et al. [61]

CCC, clear cell carcinoma; SCC, squamous cell carcinoma; ADC, adenocarcinoma; LCNEC, large cell neuroendocrine carcinoma; AdSq, adenosquamous carcinoma; HSIL, high-grade intraepithelial lesion; HPV-asso CC, HPV-associated cervical cancer; HPV-ind CC, HPV-independent cervical cancer; SCJ, squamocolumnar junction; ODX, organoid-derived xenograft; PDO, patinet-derived organoid; BME, basement membrane extract; EGF, epidermal growth factor; FGF, fibroblast growth factor; CM, conditioned medium

[a] Excluding basal media composition such as Glutamax, HEPES, and antibiotics

small materials. However, the success rates of PDO establishment from brush materials were 50% and 25% for the SCC and ADC subtypes, respectively, warranting further optimization of the applied protocol. Recently, four PDOs were established from biopsy samples of cervical cancer with various histological types, namely, SCC, ADC, large neuroendocrine carcinoma (LCNEC), and villoglandular ADC (VGA) [54]. Among the 171 compounds tested, 7 candidates exhibited high efficacy. ADC- and LCNEC-PDOs were resistant to radiation exposure, whereas SCC- and VGA-PDOs were sensitive. Additionally, 67 cervical cancer organoids have been established from 87 patients, 70% of which were SCC-derived PDOs [55]. The radiosensitivity analysis of 14 cervical cancer organoids showed significantly varying results among PDOs, representing the radiotherapy response of the patients.

Consistent with this observation, the antitumor effect of radiotherapy varied even in organoid-turned xenografts.

Recently, immunotherapy has emerged as a promising approach for treating patients with cervical cancer. A thorough elucidation of the tumor cell–microenvironment interaction is critical to better understand tumor immunity in cervical cancer. However, standard organoid cultures consist only of epithelial cells, thereby lacking such interactions. To address this issue, an in vitro tumor infiltrating lymphocyte (TIL)-specific killing assay based on the coculture of PDOs and TILs has been performed [55]. The killing efficiency of different tumor-specific TILs varied in vitro. Although the results of in vitro killing assays were not compared with the clinical outcomes, these findings demonstrate the potential of the PDO platform for in vitro prediction of adoptive cellular immunotherapy efficacy.

Long-term persistent HPV infection of the cervical epithelium causes intraepithelial lesions, including low-grade squamous intraepithelial lesions (SILs), which can develop into high-grade SIL (HSIL) and eventually into SCC. Although the premalignant state preceding neoplasia is well documented, its human-derived in vitro models were not available. Recently, a long-term organoid culture protocol for HSIL and SCC was developed that accurately recapitulated the properties of each original tissue [56]. HSIL-PDOs exhibit more sensitivity to cisplatin, carboplatin, gemcitabine, and olaparib than that by SCC-PDOs. Coculture of PDOs with HPV antigenic peptide-stimulated peripheral blood immune cells (PBMCs) showed that each PDO responded differentially to the immunized PBMCs. HSIL- and SCC-PDOs behaved differentially when cocultured with PBMCs activated by different HPV antigenic peptides. These results strongly suggest that immune cells recognize different antigenic peptide stimuli. Reportedly, the PI3K/AKT signaling pathway is highly activated in HPV-independent cervical cancer [57]. Consistent with these findings, HPV-independent PDO exhibits a lower IC50 of alpelisib, a PI3K α-selective inhibitor, than HPV-associated PDOs, suggesting the potential significance of PI3K α inhibitors in patients with HPV-independent cervical cancer.

7.3.3 *Normal Tissue-Derived Organoids*

The metaplastic epithelium of the transformation zone, including the squamocolumnar junction (SCJ) of the uterine cervix, is a primary target of HPV infection and subsequent cancer development. Organoids originating from a normal SCJ region have been propagated [58]. Many SCJ markers such as MMP7 and AGR2 exhibit higher expression in organoids than in the widely used non-SCJ cervical cell lines Ect1/E6E7 and End1/E6E7 [59]. Some organoids consist of dual cell populations: PAS^+, $p40^-$ endocervix/columnar cells and PAS^-, $p40^+$ ectocervix/squamous cells. Ecto- and endocervical organoids have also been generated from normal adult cervical tissues [53, 60]. The ectocervical organoids show a dense phenotype, whereas the endocervical organoids form cystic structures [53]. Endocervix-derived cells cultured in a WNT-deficient medium differentiate into stratified organoids resembling ectocervix organoids. Conversely, ectocervix-derived cells give rise to only squamous organoids independent of WNT [60]. These findings suggest the critical roles of the WNT signaling pathway in the cervical squamocolumnar homeostasis and the emergence of metaplasia. In contrast, the WNT pathway has been reported to play marginal roles in the trans-differentiation between the ecto- and endocervical fates [53]. Further investigations are necessary to elucidate the mechanism underlying trans-differentiation between ecto- and endocervical cells.

The potential of organoids as an in vitro model of viral infection has been evaluated by infecting ecto- and endocervical organoids with tdTomato-tagged genital HSV-1; the infected organoids collapsed [53]. Additionally, an increase in viral DNA content was observed in the organoids even at 1 week postinfection. Individual and coinfection models for infection with Chlamydia and HPV16-derived E6E7 have been reported [61]. HPV16 E6E7-expressing ectocervical organoids retained their self-renewal capacity and formed mature stratified epithelia similar to normal organoids. HPV E6E7 suppressed the intracellular development of Chlamydia and induced its persistence, whereas Chlamydia suppressed the HPV-induced mismatch repair pathway. These phenomena suggest that the effect of HPV and Chlamydia co-persistence on stem cells could be deleterious to cellular and genomic stability, thereby potentially promoting neoplastic progression. Collectively, these organoids will likely contribute to elucidating the interactions between various viral infections and epithelial cells.

7.4 Future Perspectives and Conclusions

Recent omics analyses have notably contributed to gaining insights into the mechanisms underlying carcinogenesis and tumor progression by elucidating clonal evolution in metastasis, recurrence, and drug resistance. PDOs are a promising resource, and further functional analyses are required to demonstrate causal relationships inferred from omics data. However, organoid culture presents several limitations. First, PDOs usually lack non-epithelial cells such as stromal cells, immune cells, and blood vessels. Hence, some coculture systems have been developed to address this issue. PDOs from colon cancer and cancer-associated fibroblasts have been cocultured to examine their interactions [62]. Furthermore, tumor immune responses have been evaluated by coculturing PDO and peripheral blood lymphocytes [63] or TILs [64]. These studies provide a detailed analysis of the interactions between different cell types. Second, the establishment rate of PDO varies considerably according to the cancer type and tumor grade, ranging from <20% for prostate cancer [65] to >90% for colon cancer [66]. Additionally, contamination by normal epithelial cells surrounding tumor tissues has been reported to lead to the overgrowth of normal organoids [67]. To address this issue, PDOs can be enriched by changing the composition of the culture medium for cells harboring aberrations in TP53, RB, and WNT pathway [30, 68]. Finally, organoid culture and assay protocols vary among researchers and laboratories even for the same cancer type. Such differences may affect the outcomes of the organoid-based experiments. Despite these issues, PDOs have the potential to contribute considerably to both basic and translational cancer research. We hope that this chapter encourages researchers to employ organoid culture techniques.

References

1. Azar J, Bahmad HF, Daher D, Moubarak MM, Hadadeh O, Monzer A, et al. The use of stem cell-derived organoids in disease modeling: an update. Int J Mol Sci. 2021;22(14):7667. https://doi.org/10.3390/ijms22147667.
2. Clevers H. Modeling development and disease with organoids. Cell. 2016;165(7):1586–97. https://doi.org/10.1016/j.cell.2016.05.082.
3. Koike H, Iwasawa K, Ouchi R, Maezawa M, Giesbrecht K, Saiki N, et al. Modelling human hepato-biliary-pancreatic organogenesis from the foregut-midgut boundary. Nature. 2019;574(7776):112–6. https://doi.org/10.1038/s41586-019-1598-0.
4. Sato T, Vries RG, Snippert HJ, van de Wetering M, Barker N, Stange DE, et al. Single Lgr5 stem cells build crypt-villus structures in vitro without a mesenchymal niche. Nature. 2009;459(7244):262–5. https://doi.org/10.1038/nature07935.
5. Kim J, Koo BK, Knoblich JA. Human organoids: model systems for human biology and medicine. Nat Rev Mol Cell Biol. 2020;21(10):571–84. https://doi.org/10.1038/s41580-020-0259-3.
6. Hammerhøj A, Chakravarti D, Sato T, Jensen KB, Nielsen OH. Organoids as regenerative medicine for inflammatory bowel disease. iScience. 2024;27(6):110118. https://doi.org/10.1016/j.isci.2024.110118.
7. Bartfeld S, Bayram T, van de Wetering M, Huch M, Begthel H, Kujala P, et al. In vitro expansion of human gastric epithelial stem cells and their responses to bacterial infection. Gastroenterology. 2015;148(1):126–36.e6. https://doi.org/10.1053/j.gastro.2014.09.042.
8. Drost J, Clevers H. Organoids in cancer research. Nat Rev Cancer. 2018;18(7):407–18. https://doi.org/10.1038/s41568-018-0007-6.
9. Maru Y, Onuma K, Ochiai M, Imai T, Hippo Y. Shortcuts to intestinal carcinogenesis by genetic engineering in organoids. Cancer Sci. 2019;110(3):858–66. https://doi.org/10.1111/cas.13938.
10. Maru Y, Tanaka N, Tatsumi Y, Nakamura Y, Itami M, Hippo Y. Kras activation in endometrial organoids drives cellular transformation and epithelial-mesenchymal transition. Oncogene. 2021;10(6):46. https://doi.org/10.1038/s41389-021-00337-8.
11. Maru Y, Tanaka N, Tatsumi Y, Nakamura Y, Yao R, Noda T, et al. Probing the tumorigenic potential of genetic interactions reconstituted in murine fallopian tube organoids. J Pathol. 2021;255(2):177–89. https://doi.org/10.1002/path.5752.
12. Hanahan D, Weinberg RA. Hallmarks of cancer: the next generation. Cell. 2011;144(5):646–74. https://doi.org/10.1016/j.cell.2011.02.013.
13. Vogelstein B, Papadopoulos N, Velculescu VE, Zhou S, Diaz LA Jr, Kinzler KW. Cancer genome landscapes. Science (New York, NY). 2013;339(6127):1546–58. https://doi.org/10.1126/science.1235122.
14. Dagogo-Jack I, Shaw AT. Tumour heterogeneity and resistance to cancer therapies. Nat Rev Clin Oncol. 2018;15(2):81–94. https://doi.org/10.1038/nrclinonc.2017.166.
15. Maru Y, Hippo Y. Current status of patient-derived ovarian cancer models. Cells. 2019;8(5):505. https://doi.org/10.3390/cells8050505.
16. Sato T, Stange DE, Ferrante M, Vries RG, Van Es JH, Van den Brink S, et al. Long-term expansion of epithelial organoids from human colon, adenoma, adenocarcinoma, and Barrett's epithelium. Gastroenterology. 2011;141(5):1762–72. https://doi.org/10.1053/j.gastro.2011.07.050.
17. Boj SF, Hwang CI, Baker LA, Chio II, Engle DD, Corbo V, et al. Organoid models of human and mouse ductal pancreatic cancer. Cell. 2015;160(1–2):324–38. https://doi.org/10.1016/j.cell.2014.12.021.

18. Nanki K, Toshimitsu K, Takano A, Fujii M, Shimokawa M, Ohta Y, et al. Divergent routes toward Wnt and R-spondin niche independency during human gastric carcinogenesis. Cell. 2018;174(4):856–69.e17. https://doi.org/10.1016/j.cell.2018.07.027.
19. Sachs N, de Ligt J, Kopper O, Gogola E, Bounova G, Weeber F, et al. A living biobank of breast cancer organoids captures disease heterogeneity. Cell. 2018;172(1–2):373–86.e10. https://doi.org/10.1016/j.cell.2017.11.010.
20. Lee SH, Hu W, Matulay JT, Silva MV, Owczarek TB, Kim K, et al. Tumor evolution and drug response in patient-derived organoid models of bladder cancer. Cell. 2018;173(2):515–28.e17. https://doi.org/10.1016/j.cell.2018.03.017.
21. Li X, Francies HE, Secrier M, Perner J, Miremadi A, Galeano-Dalmau N, et al. Organoid cultures recapitulate esophageal adenocarcinoma heterogeneity providing a model for clonality studies and precision therapeutics. Nat Commun. 2018;9(1):2983. https://doi.org/10.1038/s41467-018-05190-9.
22. Kim M, Mun H, Sung CO, Cho EJ, Jeon HJ, Chun SM, et al. Patient-derived lung cancer organoids as in vitro cancer models for therapeutic screening. Nat Commun. 2019;10(1):3991. https://doi.org/10.1038/s41467-019-11867-6.
23. Maru Y, Tanaka N, Itami M, Hippo Y. Efficient use of patient-derived organoids as a preclinical model for gynecologic tumors. Gynecol Oncol. 2019;154(1):189–98. https://doi.org/10.1016/j.ygyno.2019.05.005.
24. Maru Y, Tanaka N, Ebisawa K, Odaka A, Sugiyama T, Itami M, et al. Establishment and characterization of patient-derived organoids from a young patient with cervical clear cell carcinoma. Cancer Sci. 2019;110(9):2992–3005. https://doi.org/10.1111/cas.14119.
25. Vlachogiannis G, Hedayat S. Patient-derived organoids model treatment response of metastatic gastrointestinal cancers. Science (New York, NY). 2018;359(6378):920–6. https://doi.org/10.1126/science.aao2774.
26. Wensink GE, Elias SG, Mullenders J, Koopman M, Boj SF, Kranenburg OW, et al. Patient-derived organoids as a predictive biomarker for treatment response in cancer patients. NPJ Precis Oncol. 2021;5(1):30. https://doi.org/10.1038/s41698-021-00168-1.
27. Thorel L, Perréard M, Florent R, Divoux J, Coffy S, Vincent A, et al. Patient-derived tumor organoids: a new avenue for preclinical research and precision medicine in oncology. Exp Mol Med. 2024;56:1531. https://doi.org/10.1038/s12276-024-01272-5.
28. Saito Y, Muramatsu T, Kanai Y, Ojima H, Sukeda A, Hiraoka N, et al. Establishment of patient-derived organoids and drug screening for biliary tract carcinoma. Cell Rep. 2019;27(4):1265–76.e4. https://doi.org/10.1016/j.celrep.2019.03.088.
29. Yan HHN, Siu HC, Law S, Ho SL, Yue SSK, Tsui WY, et al. A comprehensive human gastric cancer organoid biobank captures tumor subtype heterogeneity and enables therapeutic screening. Cell Stem Cell. 2018;23(6):882–97.e11. https://doi.org/10.1016/j.stem.2018.09.016.
30. Kawasaki K, Toshimitsu K, Matano M, Fujita M, Fujii M, Togasaki K, et al. An organoid biobank of neuroendocrine neoplasms enables genotype-phenotype mapping. Cell. 2020;183(5):1420–35.e21. https://doi.org/10.1016/j.cell.2020.10.023.
31. Kim SC, Park JW, Seo HY, Kim M, Park JH, Kim GH, et al. Multifocal organoid capturing of colon cancer reveals pervasive intratumoral heterogenous drug responses. Adv Sci (Weinh). 2022;9(5):e2103360. https://doi.org/10.1002/advs.202103360.
32. de Witte CJ, Espejo Valle-Inclan J, Hami N, Lõhmussaar K, Kopper O, Vreuls CPH, et al. Patient-derived ovarian cancer organoids mimic clinical response and exhibit heterogeneous inter- and intrapatient drug responses. Cell Rep. 2020;31(11):107762. https://doi.org/10.1016/j.celrep.2020.107762.
33. Okamoto T, duVerle D, Yaginuma K, Natsume Y, Yamanaka H, Kusama D, et al. Comparative analysis of patient-matched PDOs revealed a reduction in OLFM4-associated clusters in metastatic lesions in colorectal cancer. Stem Cell Rep. 2021;16(4):954–67. https://doi.org/10.1016/j.stemcr.2021.02.012.
34. Siegel RL, Miller KD, Wagle NS, Jemal A. Cancer statistics, 2023. CA Cancer J Clin. 2023;73(1):17–48. https://doi.org/10.3322/caac.21763.

35. Morice P, Leary A, Creutzberg C, Abu-Rustum N, Darai E. Endometrial cancer. Lancet (London, England). 2016;387(10023):1094–108. https://doi.org/10.1016/s0140-6736(15)00130-0.
36. Murali R, Soslow RA, Weigelt B. Classification of endometrial carcinoma: more than two types. Lancet Oncol. 2014;15(7):e268–78. https://doi.org/10.1016/s1470-2045(13)70591-6.
37. Kandoth C, Schultz N, Cherniack AD, Akbani R, Liu Y, Shen H, et al. Integrated genomic characterization of endometrial carcinoma. Nature. 2013;497(7447):67–73. https://doi.org/10.1038/nature12113.
38. Lee SY. Tailored therapy based on molecular characteristics in endometrial cancer. Biomed Res Int. 2021;2021:2068023. https://doi.org/10.1155/2021/2068023.
39. Girda E, Huang EC, Leiserowitz GS, Smith LH. The use of endometrial cancer patient-derived organoid culture for drug sensitivity testing is feasible. Int J Gynecol Cancer. 2017;27(8):1701–7. https://doi.org/10.1097/igc.0000000000001061.
40. Boretto M, Maenhoudt N, Luo X, Hennes A, Boeckx B, Bui B, et al. Patient-derived organoids from endometrial disease capture clinical heterogeneity and are amenable to drug screening. Nat Cell Biol. 2019;21(8):1041–51. https://doi.org/10.1038/s41556-019-0360-z.
41. Bi J, Newtson AM, Zhang Y, Devor EJ, Samuelson MI, Thiel KW, et al. Successful patient-derived organoid culture of gynecologic cancers for disease modeling and drug sensitivity testing. Cancer. 2021;13(12):2901. https://doi.org/10.3390/cancers13122901.
42. Berg HF, Hjelmeland ME, Lien H, Espedal H, Fonnes T, Srivastava A, et al. Patient-derived organoids reflect the genetic profile of endometrial tumors and predict patient prognosis. Commun Med (Lond). 2021;1:20. https://doi.org/10.1038/s43856-021-00019-x.
43. Jamaluddin MFB, Ko YA, Ghosh A, Syed SM, Ius Y, O'Sullivan R, et al. Proteomic and functional characterization of intra-tumor heterogeneity in human endometrial cancer. Cell Rep Med. 2022;3(9):100738. https://doi.org/10.1016/j.xcrm.2022.100738.
44. Maru Y, Kohno M, Suzuka K, Odaka A, Masuda M, Araki A, et al. Establishment and characterization of multiple patient-derived organoids from a case of advanced endometrial cancer. Hum Cell. 2024;37(3):840–53. https://doi.org/10.1007/s13577-024-01048-z.
45. Boretto M, Cox B, Noben M, Hendriks N, Fassbender A, Roose H, et al. Development of organoids from mouse and human endometrium showing endometrial epithelium physiology and long-term expandability. Development. 2017;144(10):1775–86. https://doi.org/10.1242/dev.148478.
46. Turco MY, Gardner L, Hughes J, Cindrova-Davies T, Gomez MJ, Farrell L, et al. Long-term, hormone-responsive organoid cultures of human endometrium in a chemically defined medium. Nat Cell Biol. 2017;19(5):568–77. https://doi.org/10.1038/ncb3516.
47. Sahasrabuddhe VV, Parham GP, Mwanahamuntu MH, Vermund SH. Cervical cancer prevention in low- and middle-income countries: feasible, affordable, essential. Cancer Prev Res (Phila). 2012;5(1):11–7. https://doi.org/10.1158/1940-6207.Capr-11-0540.
48. Watson M, Saraiya M, Benard V, Coughlin SS, Flowers L, Cokkinides V, et al. Burden of cervical cancer in the United States, 1998-2003. Cancer. 2008;113(10 Suppl):2855–64. https://doi.org/10.1002/cncr.23756.
49. Karamurzin YS, Kiyokawa T, Parkash V, Jotwani AR, Patel P, Pike MC, et al. Gastric-type endocervical adenocarcinoma: an aggressive tumor with unusual metastatic patterns and poor prognosis. Am J Surg Pathol. 2015;39(11):1449–57. https://doi.org/10.1097/pas.0000000000000532.
50. The Cancer Genome Atlas Research Network. Integrated genomic and molecular characterization of cervical cancer. Nature. 2017;543(7645):378–84. https://doi.org/10.1038/nature21386.
51. Friedman CF, Ravichandran V, Miller K, Vanderbilt C, Zhou Q, Iasonos A, et al. Assessing the genomic landscape of cervical cancers: clinical opportunities and therapeutic targets. Clin Cancer Res. 2023;29:4660. https://doi.org/10.1158/1078-0432.Ccr-23-1078.
52. Yao J, Takenaga K, Koshikawa N, Kida Y, Lin J, Watanabe T, et al. Anticancer effect of a pyrrole-imidazole polyamide-triphenylphosphonium conjugate selectively targeting a common mitochondrial DNA cancer risk variant in cervical cancer cells. Int J Cancer. 2023;152(5):962–76. https://doi.org/10.1002/ijc.34319.

53. Lõhmussaar K, Oka R, Espejo Valle-Inclan J, Smits MHH, Wardak H, Korving J, et al. Patient-derived organoids model cervical tissue dynamics and viral oncogenesis in cervical cancer. Cell Stem Cell. 2021;28(8):1380–96.e6. https://doi.org/10.1016/j.stem.2021.03.012.
54. Seol HS, Oh JH, Choi E, Kim S, Kim H, Nam EJ. Preclinical investigation of patient-derived cervical cancer organoids for precision medicine. J Gynecol Oncol. 2023;34(3):e35. https://doi.org/10.3802/jgo.2023.34.e35.
55. Huang H, Pan Y, Huang J, Zhang C, Liao Y, Du Q, et al. Patient-derived organoids as personalized avatars and a potential immunotherapy model in cervical cancer. iScience. 2023;26(11):108198. https://doi.org/10.1016/j.isci.2023.108198.
56. Hu B, Wang R, Wu D, Long R, Fan J, Hu Z, et al. A promising new model: establishment of patient-derived organoid models covering HPV-related cervical pre-cancerous lesions and their cancers. Adv Sci (Weinh). 2024;11(12):e2302340. https://doi.org/10.1002/advs.202302340.
57. Wang Y, He M, He T, Ouyang X, Shen X, Shi W, et al. Integrated genomic and transcriptomic analysis reveals the activation of PI3K signaling pathway in HPV-independent cervical cancers. Br J Cancer. 2024;130(6):987–1000. https://doi.org/10.1038/s41416-023-02555-w.
58. Maru Y, Kawata A, Taguchi A, Ishii Y, Baba S, Mori M, et al. Establishment and molecular phenotyping of organoids from the squamocolumnar junction region of the uterine cervix. Cancer. 2020;12(3):694. https://doi.org/10.3390/cancers12030694.
59. Fichorova RN, Rheinwald JG, Anderson DJ. Generation of papillomavirus-immortalized cell lines from normal human ectocervical, endocervical, and vaginal epithelium that maintain expression of tissue-specific differentiation proteins. Biol Reprod. 1997;57(4):847–55. https://doi.org/10.1095/biolreprod57.4.847.
60. Chumduri C, Gurumurthy RK, Berger H, Dietrich O, Kumar N, Koster S, et al. Opposing Wnt signals regulate cervical squamocolumnar homeostasis and emergence of metaplasia. Nat Cell Biol. 2021;23(2):184–97. https://doi.org/10.1038/s41556-020-00619-0.
61. Koster S, Gurumurthy RK, Kumar N, Prakash PG, Dhanraj J, Bayer S, et al. Modelling chlamydia and HPV co-infection in patient-derived ectocervix organoids reveals distinct cellular reprogramming. Nat Commun. 2022;13(1):1030. https://doi.org/10.1038/s41467-022-28569-1.
62. Strating E, Verhagen MP, Wensink E, Dünnebach E, Wijler L, Aranguren I, et al. Co-cultures of colon cancer cells and cancer-associated fibroblasts recapitulate the aggressive features of mesenchymal-like colon cancer. Front Immunol. 2023;14:1053920. https://doi.org/10.3389/fimmu.2023.1053920.
63. Dijkstra KK, Cattaneo CM, Weeber F, Chalabi M, van de Haar J, Fanchi LF, et al. Generation of tumor-reactive T cells by co-culture of peripheral blood lymphocytes and tumor organoids. Cell. 2018;174(6):1586–98.e12. https://doi.org/10.1016/j.cell.2018.07.009.
64. Parikh AY, Masi R, Gasmi B, Hanada KI, Parkhurst M, Gartner J, et al. Using patient-derived tumor organoids from common epithelial cancers to analyze personalized T-cell responses to neoantigens. Cancer Immunol Immunother. 2023;72(10):3149–62. https://doi.org/10.1007/s00262-023-03476-6.
65. Pauli C, Hopkins BD, Prandi D, Shaw R, Fedrizzi T, Sboner A, et al. Personalized in vitro and in vivo cancer models to guide precision medicine. Cancer Discov. 2017;7(5):462–77. https://doi.org/10.1158/2159-8290.cd-16-1154.
66. Fujii M, Shimokawa M, Date S, Takano A, Matano M, Nanki K, et al. A colorectal tumor organoid library demonstrates progressive loss of niche factor requirements during tumorigenesis. Cell Stem Cell. 2016;18(6):827–38. https://doi.org/10.1016/j.stem.2016.04.003.
67. Dijkstra KK, Monkhorst K, Schipper LJ, Hartemink KJ, Smit EF, Kaing S, et al. Challenges in establishing pure lung cancer organoids limit their utility for personalized medicine. Cell Rep. 2020;31(5):107588. https://doi.org/10.1016/j.celrep.2020.107588.
68. Wallaschek N, Niklas C, Pompaiah M, Wiegering A, Germer CT, Kircher S, et al. Establishing pure cancer organoid cultures: identification, selection and verification of cancer phenotypes and genotypes. J Mol Biol. 2019;431(15):2884–93. https://doi.org/10.1016/j.jmb.2019.05.031.

Chapter 8
Cancer Cell Lines in the Twenty-First Century: Applications in Multi-omics and Drug Development

Tadashi Kondo

Abstract The integration of multi-omics technologies and precision oncology has transformed cancer research in the twenty-first century, demanding model systems that are both experimentally tractable and molecularly representative. Despite the emergence of complex models such as organoids and patient-derived xenografts (PDXs), cancer cell lines remain indispensable due to their scalability, genetic accessibility, and compatibility with high-throughput applications. Cell lines have played a central role in the functional interpretation of omics data, the identification of therapeutic vulnerabilities, and the development of pharmacogenomic resources such as the Cancer Dependency Map (DepMap) and ProCan-DepMapSanger. Recent advances in pharmaco-proteogenomics have further leveraged large, annotated cell line panels to uncover molecular correlates of drug response and nominate candidate biomarkers. Nonetheless, long-established cell lines suffer from limitations including genomic drift, lack of clinical annotation, and underrepresentation of rare cancer subtypes—factors that compromise reproducibility and translational utility. To address these challenges, a new generation of patient-derived cancer cell lines must be genomically validated, clinically annotated, and deposited in curated biorepositories with standardized metadata. Such efforts will reinforce the role of cell lines as foundational tools in data-driven cancer biology, ensuring their continued relevance in an era of systems oncology and precision therapeutics.

Keywords Cancer cell lines · Multi-omics · Pharmaco-proteogenomics · Drug development · Translational research

---

T. Kondo (✉)
Division of Rare Cancer Research, National Cancer Center, Chuo-ku, Tokyo, Japan

N. Shinomiya, Y. Shimada (eds.), *Applications of Cell Culture*, Current Human Cell Research and Applications, https://doi.org/10.1007/978-981-95-0587-6_8

8.1 Introduction

Patient-derived cancer models have long served as indispensable tools in cancer research, providing controlled experimental systems to investigate tumor biology, molecular mechanisms, and therapeutic responses. These models span a spectrum of complexity, from conventional cell lines to more physiologically complex platforms such as organoids and patient-derived xenografts (PDXs), and have contributed substantially to advances in functional studies for genes and proteins, validation of hypothesis, and preclinical drug development.

In the twenty-first century, cancer research has undergone a paradigm shift driven by the widespread application of high-throughput omics technologies—including genomics, transcriptomics, proteomics, and metabolomics—and the clinical implementation of molecularly targeted therapies and immune checkpoint inhibitors. These developments have been supported by the expansion of biobanking infrastructure and further accelerated by advances in digital technologies, including web-based data sharing, curated biological data repositories, and bioinformatics platforms. Together, these systems enable the generation, integration, and analysis of large-scale molecular datasets across diverse cancer types.

Within this evolving research ecosystem, patient-derived cancer models remain central. A defining advantage of these models is their capacity to respond to external perturbations, such as anticancer compounds or genetic interventions. This responsiveness enables both pharmacological profiling and functional exploration of molecular pathways. Among available model systems, cancer cell lines offer unique value due to their scalability, stability, and compatibility with high-throughput experimentation—making them ideally suited for systematic, multi-parametric investigations.

This review discusses how patient-derived cancer models, particularly cancer cell lines, are positioned to meet the current demands of omics-based and therapeutic cancer research. It further examines how these models have been integrated into large-scale molecular studies and highlights key limitations that must be addressed to improve their translational and experimental utility.

8.2 Multi-omics Studies and the Role of Cell Lines

In recent years, multilayered omics analyses—including genomics, transcriptomics, proteomics, and metabolomics—have become integral to cancer research [1, 2]. These technologies, once confined to specialized academic centers, are now supported by publicly funded research infrastructures and harmonized protocols, enabling broader accessibility and reproducibility. It is now feasible to interrogate individual malignancies from multiple molecular dimensions, offering an unprecedented view of tumor complexity [3–6]. Anticipation continues to grow that integrative omics approaches may yield a holistic understanding of cancer pathogenesis and therapeutic vulnerability, previously unattainable through conventional single-layer investigations [7]. Reflecting this shift, cancer genome medicine has gained traction in clinical practice, with genome-informed diagnostics and treatment recommendations now incorporated into national healthcare systems [8–10]. The implementation of genome medicine has also contributed to improvements in clinical trial design, particularly in patient stratification and molecular matching [11].

The expansion of omics-driven research has been closely linked to the development of biobanking and data-sharing infrastructures. The UK Biobank, for example, has amassed biological specimens and longitudinal health data from over 500,000 individuals [12, 13]. A recent large-scale study utilizing this resource applied Olink's Proximity Extension Assay (PEA) to generate plasma proteomic profiles for more than 50,000 participants, integrating them with genotypic and phenotypic information [14]. In the PEA, pairs of oligonucleotide-labeled antibodies bind to target proteins in close proximity, enabling hybridization of their DNA tags, which are then amplified and quantified by qPCR, allowing highly sensitive and multiplexed protein detection from minimal sample volume. This work identified thousands of protein quantitative trait loci (pQTLs), with broad implications for biomarker discovery and therapeutic targeting. Another study also used the UK biobank resource and Olink's assay system in the biomarker study of multiple sclerosis risk and reported protein biomarkers of disease severity [15]. In Japan, the NCC Cancer Biobank has collected tumor tissues, blood samples, and detailed clinical information from patients across diverse cancer types [16]. The biospecimens from various hospitals and biobanks have supported nationwide multi-omics projects such as MONSTAR-SCREEN and SCRUM-Japan GI-SCREEN, which link genomic and transcriptomic data with pharmacologic profiles to guide precision oncology [17, 18]. These initiatives have further facilitated the social implementation of genome medicine by supporting molecularly guided clinical trial enrollment for patients with advanced cancers [19].

Parallel efforts to centralize molecular datasets have also transformed hypothesis generation. The Cancer Genome Atlas (TCGA), for instance, has archived genomic, epigenomic, transcriptomic, and proteomic profiles from over 11,000 tumor samples across 33 cancer types [20]. The large-scale data from cancer genomic studies such as The Cancer Genome Atlas (TCGA), Human Cancer Models Initiative (HCMI), Foundation Medicine Inc. (FMI), and Clinical Proteomic Tumor Analysis Consortium (CPTAC) are available from Genome Data Common (GDC) Data Portal [21]. The Human Protein Atlas (HPA) integrates data from various omics technologies, including antibody-based immunohistochemistry, transcriptomics, and mass spectrometry-based proteomics [22]. These openly accessible resources provide a foundation for identifying new molecular subtypes, elucidating oncogenic mechanisms, and nominating therapeutic targets.

Patient-derived cancer models are indispensable for translating such high-dimensional data into biological insight. Multi-omics analyses routinely yield candidate genes, proteins, and pathways that may underlie disease progression or drug response. However, experimental validation is essential to distinguish functionally relevant alterations from background variability. At the hypothesis-generation stage, the volume of molecular candidates is typically too large for focused low-throughput assays. Thus, scalable and reproducible model systems are needed for initial screening.

Cancer cell lines are ideally suited to this task. Their ease of culture, amenability to genetic perturbation, and compatibility with drug screening make them powerful tools for testing omics-derived hypotheses. When a molecular alteration is found to influence phenotype in vitro, it can then be prioritized for follow-up in more physiologically relevant systems.

Organoid and patient-derived xenograft (PDX) models have been proposed as next-tier platforms that better preserve tumor architecture and microenvironmental context. While promising, these models are more resource-intensive and less standardized, and their predictive validity for clinical drug response remains under investigation [23, 24]. Therefore, cancer cell lines are widely regarded as the most accessible and scalable model systems for functionally interpreting omics data in the context of translational cancer research.

8.3 Anticancer Drug Development and the Role of Cell Lines

Over the past two decades, the development of molecularly targeted therapies has transformed the treatment landscape of several malignancies. Imatinib (Glivec®), initially approved for chronic myeloid leukemia (CML), was later shown to be effective in gastrointestinal stromal tumors (GIST), representing a paradigm shift in mechanism-based therapy [25]. Gefitinib (Iressa®), an epidermal growth factor receptor (EGFR) tyrosine kinase inhibitor, marked another milestone, particularly in non-small cell lung cancer (NSCLC). Its efficacy was first demonstrated in the IDEAL 1 and IDEAL 2 trials [26] and further validated in the IPASS trial, which established EGFR mutation status as a predictive biomarker for gefitinib response [27]. These early examples laid the foundation for molecularly stratified drug development based on oncogenic driver alterations.

Since then, an increasing number of targeted agents and immune checkpoint inhibitors have entered clinical use. As of 2025, more than 100 molecularly targeted therapies and at least 8 immune-based therapies have been approved by the US Food and Drug Administration (FDA) for the treatment of various cancers [28]. While many of these agents have become standard treatments for malignancies such as NSCLC and melanoma, their applicability across other cancer types remains uneven. Rare cancers, in particular, suffer from limited access to clinical trials due to small patient populations [29, 30]. Even for common cancers, the development of new agents remains costly, slow, and inefficient, with low overall trial success rates [31]. This reality underscores the need for robust preclinical filtering strategies to prioritize promising compounds and identify responsive molecular subtypes before advancing to clinical trials. Yet such strategies remain inconsistently applied. Consequently, many agents fail in late-phase development, driving up the cost of drug development and, ultimately, the price of newly approved therapies [32, 33].

To mitigate this, well-validated preclinical models are essential for assessing compound efficacy, identifying predictive biomarkers, and characterizing the molecular basis of drug response.

Among patient-derived cancer models, cell lines are particularly well suited to these tasks. Their scalability, genetic tractability, and compatibility with high-throughput screening platforms make them ideal for testing candidate agents across panels representing diverse tissue types and molecular subtypes. This approach enables the identification of subtype-specific drug sensitivities, potential resistance mechanisms, and repositioning opportunities for existing agents. Cell lines are thus integral to early-phase therapeutic triage.

The emergence of large-scale pharmacogenomic resources has further advanced this approach. The Cancer Cell Line Encyclopedia (CCLE) [34], the Genomics of Drug Sensitivity in Cancer (GDSC) [35], and the Cancer Therapeutics Response Portal (CTRP) [36] provide integrated datasets that link drug sensitivity with genomic, transcriptomic, and lineage-specific features across hundreds of cancer cell lines. These resources have enabled the discovery of predictive biomarkers and informed therapeutic stratification. For example, Iorio et al. demonstrated that ERBB2-amplified cell lines show enhanced sensitivity to lapatinib, supporting its potential utility beyond HER2-positive breast cancer [37]. High-throughput combination screens using the cell lines have uncovered synergistic interactions that help overcome resistance to monotherapies. Notably, Bashi et al. reported that 109 anti-cancer drug combinations, derived from AstraZeneca's portfolio of oncology-focused small molecules, were systematically screened across 755 human cancer cell lines. Employing a 7 × 7 concentration matrix design, the screening yielded over 4 million sensitivity data points, enabling the identification of drug combinations with potential clinical relevance [38].

Despite their utility, cancer cell lines are not without limitations. As simplified, two-dimensional cultures, they do not fully capture tumor heterogeneity, the micro-environment, or the influence of stromal and immune components. Moreover, drug response data derived from cell lines do not always translate into clinical outcomes at the individual patient level. These shortcomings have prompted the development of more physiologically representative models, such as PDXs and organoids. Organoid technologies have progressed rapidly and are increasingly compatible with pharmacologic studies. However, neither organoids nor PDXs have been definitively shown to outperform cell lines in predicting therapeutic efficacy. PDX models often rely on subcutaneous implantation into immunodeficient mice, limiting their ability to replicate metastatic behavior [39]. Additionally, stromal components of human origin are gradually replaced by murine cells over serial passages, potentially altering drug responses [40, 41]. While frequently described as more "physiologically relevant," the empirical basis for this assumption remains underdeveloped.

Given these constraints, cancer cell lines remain the most accessible and experimentally versatile platform for early-phase drug discovery. Their long-standing availability through public biorepositories and compatibility with scalable assays continue to make them a cornerstone of translational oncology and pharmacogenomics research.

8.4 Pharmaco-proteogenomics and the Role of Cancer Cell Lines

Pharmaco-proteogenomics is an emerging integrative strategy that combines multi-layered omics data—including genomics, transcriptomics, and proteomics—with pharmacologic profiling to elucidate the molecular determinants of drug response. This approach involves the systematic evaluation of patient-derived cancer models, particularly cancer cell lines and organoids, treated with investigational or clinically approved compounds. Comprehensive molecular characterization—via next-generation sequencing, mass spectrometry-based proteomics, and functional perturbation methods such as CRISPR/Cas9—enables researchers to link specific molecular features to therapeutic vulnerabilities. The overarching goals are three-fold: to identify compounds with clinical potential, to clarify the molecular basis of drug efficacy and resistance, and to discover predictive biomarkers for patient stratification. Pharmaco-proteogenomics thus offers a scalable framework for hypothesis generation, functional validation, and therapeutic prioritization in precision oncology.

A recent study applying this methodology to 65 human liver cancer organoids exemplifies its potential. By integrating pharmacologic responses with multi-omics profiles, the authors identified molecular correlates of drug sensitivity and candidate biomarkers for therapeutic guidance [42]. Such studies highlight the power of proteogenomic integration in uncovering clinically actionable insights.

One of the most comprehensive platforms enabling pharmaco-proteogenomic research is the Cancer Dependency Map (DepMap), led by the Broad Institute [43, 44]. DepMap incorporates over 1000 cancer cell lines from diverse histological backgrounds, sourced from the CCLE and other academic and commercial biorepositories. These lines undergo large-scale CRISPR-based loss-of-function screening, transcriptomic profiling, and—in many cases—proteomic and pharmacologic assays. The resulting datasets reveal lineage-specific gene dependencies, synthetic lethal interactions, and candidate drug targets across cancer types. Importantly, DepMap emphasizes data reproducibility through standardized workflows and extensive metadata. Its datasets are publicly accessible via an interactive web portal (https://depmap.org/), serving as a key resource for the global cancer research community.

A complementary advance was achieved by Gonçalves et al., who constructed a pan-cancer proteomic map of 949 human cancer cell lines [45]. This study profiled over 8000 proteins across 28 cancer types using mass spectrometry-based quantification, integrating the resulting proteomic data with corresponding genomic, transcriptomic, and pharmacologic information. Notably, proteomic profiles were found to predict drug sensitivity and gene essentiality as effectively as—or in some cases better than—transcriptomic data alone. The resulting ProCan-DepMapSanger dataset has been incorporated into the Cell Model Passports database (https://cellmodelpassports.sanger.ac.uk), offering an open-access platform for interrogating drug response mechanisms, pathway activation states, and lineage-specific vulnerabilities.

Together, DepMap and the pan-cancer proteomic initiative underscore the enduring importance of cancer cell lines in modern systems biology and translational pharmacology. These models combine scalability, reproducibility, and molecular tractability, making them uniquely suited for pharmaco-omics research. Their availability through established biorepositories—such as the American Type Culture Collection (ATCC) [46], the Leibniz Institute—German Collection of Microorganisms and Cell Cultures GmBH (DSMZ) [47], the Japanese Cancer Research Resources Bank (JCRB) [48], and CCLE [34]—has facilitated international collaboration and data harmonization. Decades of accumulated knowledge, paired with recent innovations in multi-omics integration, continue to position cell lines as essential infrastructure for data-driven drug discovery and biomarker development in precision oncology.

8.5 Current Limitations of Cancer Cell Lines in Omics-Based and Drug Development Research

Cancer cell lines have served as foundational models in molecular oncology and preclinical drug discovery for decades. However, several intrinsic limitations constrain their continued utility in the context of modern omics-driven and therapeutic research.

One of the most pressing issues is the imbalance in cancer-type representation. Publicly available cell line repositories are dominated by models derived from common malignancies, whereas rare tumors remain substantially underrepresented. For example, DepMap includes more than 2000 cancer cell lines, yet only 89 and 86 are derived from bone and soft tissue sarcomas, respectively [49]. Similarly, the pan-cancer proteomic atlas published by Gonçalves et al. profiled 949 cell lines, of which only 40 and 22 represented bone and soft tissue sarcomas, respectively [45]. Given that the sarcoma family encompasses over 100 histologically and molecularly distinct subtypes [50], these figures underscore a critical gap. As a result, large-scale omics and pharmacogenomic studies based on these datasets frequently fail to yield generalizable insights for sarcomas and other rare cancers. This limitation reflects a broader systemic challenge: the lack of representative models restricts the inclusion of rare malignancies in translational and therapeutic development pipelines.

Importantly, this issue is not limited to rare cancers. In earlier eras, when therapeutic approaches were largely cytotoxic and histology-guided, a handful of representative lines per tumor type was often considered sufficient. In the current landscape, shaped by precision oncology and the widespread implementation of cancer genome medicine, cell lines are expected not only to reflect histological subtypes but also to represent the spectrum of actionable genomic alterations. Even in well-studied cancers such as lung or colorectal carcinoma, gaps remain in the availability of cell lines harboring specific driver mutations. This has revealed a conceptual shift: so-called common cancers are increasingly viewed as composites of rare subtypes with distinct molecular backgrounds but similar clinical manifestations [51]. Moving forward, cell line development strategies must emphasize the generation of models that are genomically defined, in addition to being histologically annotated.

A second concern relates to molecular fidelity and reproducibility. A key advantage of cancer cell lines is their potential for standardized use across laboratories, enabling data integration and cross-study validation. However, many widely used lines were established decades ago, and prolonged in vitro passaging can lead to genetic drift, epigenetic remodeling, and phenotypic divergence from the original tumor. These changes can compromise experimental reproducibility, particularly in studies focused on drug sensitivity or molecular dependency mapping. While major cell banks maintain early-passage vials under rigorously controlled conditions, researchers often work with extensively passaged derivatives that may no longer accurately reflect the original tumor's biology. Misidentification and cross-contamination of cell lines also remain nontrivial problems [52–54]. Short tandem repeat (STR) profiling has become the standard for identity verification [55], but molecular authentication should go further. A growing priority is the systematic characterization of molecular features—genomic, transcriptomic, and proteomic—linked to biobanked cell stocks. This would allow researchers to monitor lineage fidelity over time and choose models that best match their experimental goals. The importance of these quality control measures is amplified by growing concern over the reproducibility of biomedical research. In a widely cited survey of over 1500 scientists, more than 70% reported failing to reproduce another laboratory's experiments, and over 50% reported reproducibility issues with their own findings [56]. For cell lines that have been in use for many years, it is likely necessary to implement more stringent authentication procedures beyond STR profiling to ensure that they retain the same molecular background.

A third limitation is the widespread lack of linkage between cancer cell lines and the clinical or pathological information of the donor patients. Historically, this disconnect mattered less when the primary aim of cell line-based research was the exploration of basic tumor biology. However, in the current era of biomarker-driven therapeutic development, such annotation has become essential. It is now well established that not all patients with the same histological cancer subtype respond similarly to a given treatment. Even among cell lines derived from the same tissue of origin, differences in clinical background—such as disease stage, prior therapies, or molecular classification—may account for divergent experimental outcomes. Without access to such data, the translational value of findings derived from these models is significantly diminished. Furthermore, the availability of multiple cell lines for a given cancer type does not guarantee comprehensive representation of clinical diversity. In the absence of detailed annotations, important subsets of patients—such as those with aggressive phenotypes, unusual metastatic patterns, or treatment-resistant disease—may be underrepresented. As researchers increasingly rely on in vitro models to identify predictors of therapeutic response, embedding clinical and pathological metadata into cell line repositories will be critical for ensuring the applicability of preclinical findings to real-world patient populations.

Collectively, these limitations point to the urgent need for the development of new, genomically validated and clinically annotated patient-derived cancer cell lines. Such models should be established with an explicit focus on molecular diversity, linked to standardized omics profiles, and accompanied by well-documented passage histories. Depositing them in curated public biobanks with full metadata—including STR authentication, donor clinical history, pathology reports, and multi-omics data—will be vital to maximize their utility. Only through such rigorous characterization and standardization can cancer cell lines maintain their central role in next-generation translational and systems oncology.

8.6 Conclusion

In the twenty-first century, cancer research has entered an era defined by multi-omics integration, data-intensive biology, and the clinical imperative of precision medicine. This transition—from reductionist frameworks to systems-level approaches—has redefined our understanding of cancer as a heterogeneous and stratified disease. At the same time, the development of molecularly targeted agents and immunotherapies has reshaped drug discovery into a biomarker-driven process that depends on models capable of capturing both biological complexity and clinical relevance.

In this evolving landscape, patient-derived cancer models remain indispensable. Among them, cancer cell lines continue to serve as the most scalable, accessible, and experimentally tractable tools for both hypothesis generation and functional validation. Their compatibility with genome editing, transcriptomic and proteomic profiling, and high-throughput pharmacological screening has positioned them at the core of contemporary translational research. While organoids and PDXs offer greater physiological fidelity in certain contexts, these models remain limited by cost, standardization, and scalability. As a result, cancer cell lines persist as the foundational in vitro systems for early-phase drug development and omics-guided discovery.

Nevertheless, the limitations of cell lines have become increasingly evident. The majority of available models are derived from common epithelial cancers, leaving many rare or histologically distinct tumor types underrepresented. Many widely used lines were established decades ago, often without comprehensive molecular annotation, and are susceptible to genetic drift and phenotypic alteration through repeated passaging. Moreover, the lack of linkage between cell lines and the clinical or pathological context of their donor patients hinders efforts to translate in vitro findings into meaningful therapeutic strategies.

To maintain and extend their utility, future efforts must prioritize the establishment of new cancer cell lines that are genetically validated, clinically annotated, and longitudinally monitored for molecular stability. These models should be deposited into rigorously curated public repositories, accompanied by standardized authentication, passage histories, diagnostic information, and multi-omics profiles. Such infrastructure would enhance cross-study comparability, improve reproducibility, and elevate the translational relevance of preclinical findings.

Cancer cell lines—if developed and curated with precision—will remain essential assets in the next generation of oncology research. Their enduring value lies not only in their biological versatility but also in their potential to serve as integrative platforms that connect molecular discovery with clinical application. As the field continues to evolve, thoughtful investment in the refinement and contextualization of cell line resources will be critical to meeting the scientific and therapeutic challenges of modern cancer medicine.

References

1. Akhoundova D, Rubin MA. Clinical application of advanced multi-omics tumor profiling: shaping precision oncology of the future. Cancer Cell. 2022;40:920–38.
2. Athieniti E, Spyrou GM. A guide to multi-omics data collection and integration for translational medicine. Comput Struct Biotechnol J. 2023;21:134–49.
3. Wang C, Li J, Chen J, et al. Multi-omics analyses reveal biological and clinical insights in recurrent stage I non-small cell lung cancer. Nat Commun. 2025;16:1477.
4. Jayakrishnan TT, Sangwan N, Barot SV, et al. Multi-omics machine learning to study host-microbiome interactions in early-onset colorectal cancer. NPJ Precis Oncol. 2024;8:146.
5. Requena D, Medico JA, Soto-Ugaldi LF, et al. Liver cancer multiomics reveals diverse protein kinase A disruptions convergently produce fibrolamellar hepatocellular carcinoma. Nat Commun. 2024;15:10887.
6. Li Y, Sun H. Multi-omics analysis identifies novels genes involved in glioma prognosis. Sci Rep. 2025;15:5806.
7. Wekesa JS, Kimwele M. A review of multi-omics data integration through deep learning approaches for disease diagnosis, prognosis, and treatment. Front Genet. 2023;14:1199087.
8. Mano H. Cancer genomic medicine in Japan. Proc Jpn Acad Ser B Phys Biol Sci. 2020;96:316–21.
9. Mukai Y, Ueno H. Establishment and implementation of cancer genomic medicine in Japan. Cancer Sci. 2021;112:970–7.
10. Kohno T, Kato M, Kohsaka S, et al. C-CAT: the national datacenter for cancer genomic medicine in Japan. Cancer Discov. 2022;12:2509–15.
11. Okuma HS, Yonemori K, Narita SN, et al. MASTER KEY Project: powering clinical development for rare cancers through a platform trial. Clin Pharmacol Ther. 2020;108:596–605.
12. Sudlow C, Gallacher J, Allen N, et al. UK Biobank: an open access resource for identifying the causes of a wide range of complex diseases of middle and old age. PLoS Med. 2015;12:e1001779.
13. Bycroft C, Freeman C, Petkova D, et al. The UK Biobank resource with deep phenotyping and genomic data. Nature. 2018;562:203–9.
14. Xu H, Li J, McClements DJ, et al. Eggshell waste act as multifunctional fillers overcoming the restrictions of starch-based films. Int J Biol Macromol. 2023;253:127165.
15. Jacobs BM, Vickaryous N, Giovannoni G, Proitsi P, Waters S, Dobson R. Plasma proteomic profiles of UK Biobank participants with multiple sclerosis. Ann Clin Transl Neurol. 2024;11:698–709.

16. Yoshida T, Yatabe Y, Kato K, et al. The evolution of cancer genomic medicine in Japan and the role of the National Cancer Center Japan. Cancer Biol Med. 2023;21:29–44.
17. Nakamura Y, Fujisawa T, Taniguchi H, et al. SCRUM-Japan GI-SCREEN and MONSTAR-SCREEN: path to the realization of biomarker-guided precision oncology in advanced solid tumors. Cancer Sci. 2021;112:4425–32.
18. Hashimoto T, Nakamura Y, Fujisawa T, et al. The SCRUM-MONSTAR cancer-omics ecosystem: striving for a quantum leap in precision medicine. Cancer Discov. 2024;14:2243–61.
19. Shiota M, Matsubara N, Kato T, et al. Genomic profiling and clinical utility of circulating tumor DNA in metastatic prostate cancer: SCRUM-Japan MONSTAR SCREEN project. BJC Rep. 2024;2:28.
20. The Cancer Genome Atlas Program (TCGA). https://www.cancer.gov/ccg/research/genome-sequencing/tcga?utm_source=chatgpt.com.
21. Genomic Data Commons. https://portal.gdc.cancer.gov/.
22. The Human Protein Atlas. https://www.proteinatlas.org/.
23. Abdolahi S, Ghazvinian Z, Muhammadnejad S, Saleh M, Asadzadeh Aghdaei H, Baghaei K. Patient-derived xenograft (PDX) models, applications and challenges in cancer research. J Transl Med. 2022;20:206.
24. Jiang X, Oyang L, Peng Q, et al. Organoids: opportunities and challenges of cancer therapy. Front Cell Dev Biol. 2023;11:1232528.
25. Cohen MH, Moses ML, Pazdur R. Gleevec for the treatment of chronic myelogenous leukemia: US. Food and Drug Administration regulatory mechanisms, accelerated approval, and orphan drug status. Oncologist. 2002;7:390–2.
26. Fukuoka M, Yano S, Giaccone G, et al. Multi-institutional randomized phase II trial of gefitinib for previously treated patients with advanced non-small-cell lung cancer (The IDEAL 1 Trial) [corrected]. J Clin Oncol. 2003;21:2237–46.
27. Maemondo M, Inoue A, Kobayashi K, et al. Gefitinib or chemotherapy for non-small-cell lung cancer with mutated EGFR. N Engl J Med. 2010;362:2380–8.
28. Drug Approvals and Databases. https://www.fda.gov/drugs/development-approval-process-drugs/drug-approvals-and-databases.
29. Pizzamiglio C, Vernon HJ, Hanna MG, Pitceathly RDS. Designing clinical trials for rare diseases: unique challenges and opportunities. Nat Rev Methods Primers. 2022:2.
30. Subbiah V, Othus M, Palma J, Cuglievan B, Kurzrock R. Designing clinical trials for patients with rare cancers: connecting the zebras. Am Soc Clin Oncol Educ Book. 2025;45:e100051.
31. Scannell JW, Blanckley A, Boldon H, Warrington B. Diagnosing the decline in pharmaceutical R&D efficiency. Nat Rev Drug Discov. 2012;11:191–200.
32. Haslam A, Olivier T, Powell K, Tuia J, Prasad V. Eventual success rate and predictors of success for oncology drugs tested in phase I trials. Int J Cancer. 2023;152:276–82.
33. Thomas D, Chancellor D, Micklus A, et al. Clinical development success rates and contributing factors 2011–2020. BIO, Informa Pharma Intelligence, QLS Advisors; 2021.
34. Cancer Cell Line Encyclopedia (CCLE). https://sites.broadinstitute.org/ccle.
35. Genomics of Drug Sensitivity in Cancer (GDSC). https://www.cancerrxgene.org/.
36. Cancer Therapeutics Response Portal (CTRP). https://portals.broadinstitute.org/ctrp.v2/.
37. Iorio F, Knijnenburg TA, Vis DJ, et al. A landscape of pharmacogenomic interactions in cancer. Cell. 2016;166:740–54.
38. Bashi AC, Coker EA, Bulusu KC, et al. Large-scale pan-cancer cell line screening identifies actionable and effective drug combinations. Cancer Discov. 2024;14:846–65.
39. De Angelis ML, Francescangeli F, Nicolazzo C, et al. An orthotopic patient-derived xenograft (PDX) model allows the analysis of metastasis-associated features in colorectal cancer. Front Oncol. 2022;12:869485.
40. Chao C, Widen SG, Wood TG, et al. Patient-derived xenografts from colorectal carcinoma: a temporal and hierarchical study of murine stromal cell replacement. Anticancer Res. 2017;37:3405–12.

41. Tumor-stromal cell turnover in PDX mice. https://www.jax.org/news-and-insights/2014/november/tumor-stromal-cell-turnover-in-pdx-mice.
42. Ji S, Feng L, Fu Z, et al. Pharmaco-proteogenomic characterization of liver cancer organoids for precision oncology. Sci Transl Med. 2023;15:eadg3358.
43. Tsherniak A, Vazquez F, Montgomery PG, et al. Defining a cancer dependency map. Cell. 2017;170:564–76.e16.
44. Arafeh R, Shibue T, Dempster JM, Hahn WC, Vazquez F. The present and future of the cancer dependency map. Nat Rev Cancer. 2024;25:59.
45. Gonçalves E, Poulos RC, Cai Z, et al. Pan-cancer proteomic map of 949 human cell lines. Cancer Cell. 2022;40:835–49.e8.
46. American Type Culture Collection (ATCC). https://www.atcc.org/.
47. Leibniz Institute—German Collection of Microorganisms and Cell Cultures GmBH (DSMZ). https://www.dsmz.de/.
48. Japanese Cancer Research Resources Bank (JCRB). https://cellbank.nibn.go.jp/english.
49. DepMap Portal. https://depmap.org/portal/.
50. WHO classification of tumours of soft tissue and bone; 2020.
51. Dieci MV, Orvieto E, Dominici M, Conte P, Guarneri V. Rare breast cancer subtypes: histological, molecular, and clinical peculiarities. Oncologist. 2014;19:805–13.
52. Register of misidentified cell lines. https://iclac.org/databases/cross-contaminations/.
53. Huang Y, Liu Y, Zheng C, Shen C. Investigation of cross-contamination and misidentification of 278 widely used tumor cell lines. PLoS One. 2017;12:e0170384.
54. Horbach S, Halffman W. The ghosts of HeLa: how cell line misidentification contaminates the scientific literature. PLoS One. 2017;12:e0186281.
55. Almeida JL, Korch CT. Authentication of human and mouse cell lines by short tandem repeat (STR) DNA genotype analysis. In: Markossian S, Grossman A, Baskir H, et al., editors. Assay guidance manual. Bethesda, MD: Eli Lilly & Company and the National Center for Advancing Translational Sciences; 2004.
56. Baker M. 1,500 scientists lift the lid on reproducibility. Nature. 2016;533:452–4.

Chapter 9
Cell Lines in Rare Cancer Research

Takashi Kondo and Tadashi Kondo

Abstract Rare cancers, while individually uncommon, account for over 20% of all cancer diagnoses worldwide and present distinct clinical and research challenges. This chapter explores the critical role of cancer cell lines in advancing rare cancer research, focusing on their utility as scalable, experimentally tractable platforms for molecular and pharmacological studies. Despite the rise of advanced models such as organoids and xenografts, conventional cell lines remain central to high-throughput investigations and multi-omics integration. Using osteosarcoma as a representative example, we analyze 226 cell lines catalogued in the Cellosaurus database. Our findings reveal that only a minority are both published in peer-reviewed literature and available from public repositories. Moreover, accessible models exhibit significant demographic and clinical biases, including underrepresentation of older patients and metastatic lesions—limiting their translational applicability. We highlight the importance of formally publishing newly established cell lines to ensure reproducibility, traceability, and community-wide use. While comprehensive characterization is ideal, excessive demands for omics data at the time of reporting may deter publication. We advocate for a balanced approach that ensures minimally sufficient documentation—such as donor metadata, STR profiling, and contamination testing—while allowing for future functional elaboration. To overcome systemic barriers, we propose structural reforms including clear editorial guidelines, publication incentives, and support for public cell banking. By fostering a culture of transparency and accessibility, the cancer research community can build a more inclusive and sustainable infrastructure to accelerate progress in the study of rare malignancies.

Keywords Rare cancers · Patient-derived cell lines · Osteosarcoma · Translational oncology · Pharmaco-proteogenomics

T. Kondo
Mito Kyodo General Hospital, University of Tsukuba, Mito, Ibaraki, Japan

T. Kondo (✉)
Division of Rare Cancer Research, National Cancer Center, Chuo-ku, Tokyo, Japan

N. Shinomiya, Y. Shimada (eds.), *Applications of Cell Culture*, Current Human Cell Research and Applications, https://doi.org/10.1007/978-981-95-0587-6_9

9.1 Introduction: Rare Cancers in a Global Context

Rare cancers are conventionally defined by low incidence, but their scientific and clinical importance belies the term "rare." In Japan, rare cancers are defined as malignancies with an incidence of fewer than 6 cases per 100,000 population per year—characterized not only by their statistical infrequency but also by the disproportionately greater clinical and healthcare challenges they present due to their rarity [1]. This definition is notable for including not only statistical rarity but also the disproportionate difficulties in diagnosis, treatment, and access to appropriate care that arise from low prevalence. Conversely, this definition implies that cancers with a low incidence are not classified as rare in Japan if their rarity does not translate into clinical or healthcare challenges. For example, tumors harboring ALK fusions or BRAF mutations in lung cancers are relatively uncommon [2, 3]; however, because they fall within the purview of established diagnostic and therapeutic frameworks managed by specialists, they are not designated as rare cancers. This incidence threshold aligns closely with the European Union's definition, which also classifies malignancies occurring in fewer than 6 individuals per 100,000 per year as rare [4]. It is also the same as the definition in Australia; rare cancers are defined as those with an incidence of less than 6 per 100,000 Australians per annum and less common cancers as those with an incidence of between 6 and 12 (inclusive) per 100,000 Australians per annum [5]. By contrast, the United States adopts a broader definition through the Orphan Drug Act, designating rare cancers as those affecting fewer than 200,000 individuals annually—approximately translating to fewer than 40 cases per 100,000 people per year [6]. This framing reflects a regulatory environment focused on incentivizing pharmaceutical development for underserved patient populations, and it allows for inclusion of cancers that may not meet strict epidemiological rarity but are still disadvantaged in terms of treatment innovation. As such, the definition of rare cancers is shaped by prevailing medical paradigms and social healthcare systems. While cancers are typically classified according to their tissue of origin or underlying molecular alterations, rare cancers are uniquely defined by the number of cases, highlighting a sociomedical dimension that sets them apart.

Despite the differences in formal thresholds and policy contexts, rare cancers present a shared global reality. While each subtype is individually uncommon, rare cancers as a group encompass over 200 distinct entities and collectively account for 20% or more of all cancer diagnoses in multiple regions [4, 7, 8]. From a public health perspective, rare cancers are not rare. Notably, when considered collectively, the total number of patients with rare cancers exceeds that of major cancer types such as lung or colorectal cancer. Thus, advancing the clinical care and research of rare cancers is a matter of national public health significance. Although rare cancer research may appear, at first glance, to be a niche field focused on small patient populations, it in fact constitutes a critically important domain within the broader landscape of oncology.

Although the total number of rare cancer cases is substantial when aggregated, each individual rare cancer remains limited in incidence, giving rise to a distinct set of challenges stemming from their rarity. For instance, rare cancers are often associated with delayed or inaccurate diagnosis [6, 9]. The paucity of cases also poses significant barriers to conducting adequately powered clinical trials, and the small target populations result in limited market potential, reducing commercial incentives for pharmaceutical investment [10, 11]. These compounding factors contribute to the generally poorer prognoses observed in rare cancers compared to more common malignancies [12, 13]. Moreover, patients with rare cancers frequently experience considerable psychological burdens related to diagnostic uncertainty, limited treatment options, social isolation, and insufficient access to information [14], underscoring the need for enhanced medical and psychosocial support. Taken together, the development of accurate diagnostic strategies and effective therapies for rare cancers represents an urgent and critical priority in contemporary oncology. To address these challenges, dedicated rare cancer centers have been established, and cross-institutional networks for research and clinical care are being developed across multiple expert facilities [15, 16]. In parallel, national and international data-sharing infrastructures are being implemented, alongside the design of academically led basket and umbrella trials tailored for patients with rare cancers—initiatives that directly confront the unique hurdles posed by low-incidence malignancies [17]. While promising, these approaches require extensive preclinical data and robust molecular characterization, both of which are often lacking for rare cancers due to limited biospecimen availability and underrepresentation in genomic studies.

This chapter examines the strategic utility of cancer cell lines in rare cancer research. Building on the broader discussion presented in Chap. 8, which surveyed the applications of cell lines across oncology, we now focus specifically on their indispensable role in modeling rare malignancies—where they often serve as the sole experimental proxy for translational investigation.

9.2 Cell Lines in the Twenty-First Century

Patient-derived cancer models including cell lines have long stood as indispensable tools in oncology research, offering reproducible, scalable, and experimentally tractable platforms to dissect the molecular underpinnings of tumor biology. Their relevance has intensified in the twenty-first century, propelled by the rapid expansion of high-throughput multi-omics technologies—including genomics, transcriptomics, proteomics, and metabolomics—which demand robust model systems for functional validation and mechanistic interrogation [18]. The capacity of cell lines to support genome-wide perturbation studies, pharmacological screening, and biomarker discovery remains unparalleled in its efficiency and cost-effectiveness. Patient-derived cancer models, such as cell lines, organoids, and xenografts, offer distinct characteristics and advantages. Notably, cell lines are particularly well-suited for high-throughput perturbation studies and large-scale screening experiments. In the context of multi-omics research, cell lines remain indispensable due to their scalability and tractability. These attributes underscore the continued relevance of cell lines as a central patient-derived model in twenty-first century cancer research. On the other hand, no patient-derived cancer model—be it a cell line, organoid, or xenograft—perfectly recapitulates the complexity of tumors in vivo. As the statistician George Box famously stated, "all models are wrong, but some are useful," arguing that while no model is entirely accurate, simplified representations can nonetheless yield valuable insights when applied appropriately [19]. This principle is particularly relevant to patient-derived cancer models such as cell lines, which, despite their limitations, remain powerful tools when used with a clear understanding of their strengths and constraints.

Among their most transformative applications has been the generation of large-scale, integrative datasets derived from molecularly annotated cell line panels. With the application of multilayered omics analyses, a clearer understanding is emerging of how and to what extent specific cell lines resemble tumors in vivo [20]. This growing body of knowledge is expected to inform more refined and context-appropriate uses of cell lines in cancer research. Landmark initiatives such as the Cancer Cell Line Encyclopedia (CCLE) [21–23], the Genomics of Drug Sensitivity in Cancer (GDSC) [24], and the Cancer Dependency Map (DepMap) [25, 26] have collectively harnessed hundreds of annotated cell lines to systematically chart oncogenic dependencies, lineage-specific vulnerabilities, and pharmacological response profiles. These efforts have catalyzed the reclassification of tumors based on molecular features, revealed novel disease taxonomies, and informed the development of mechanism-based therapeutic strategies.

The role of cell lines has further evolved with the emergence of pharmacoproteogenomics, an integrative discipline that correlates drug response data with proteomic, transcriptomic, and genomic signatures [27, 28]. By aligning compound sensitivity profiles with molecular determinants of action, this approach facilitates the identification of actionable targets, elucidates resistance mechanisms, and enables the rational design of combination therapies. Cell lines remain central to this framework, offering the experimental flexibility necessary for iterative hypothesis testing and mechanistic validation.

9.3 Cell Lines of Rare Cancer

The establishment, characterization, and dissemination of cancer cell lines are underpinned by long-standing infrastructures that enable standardized research across institutions and international borders. Once established, these models are typically characterized and reported in publications, after which many are deposited in public cell banks to ensure accessibility, reproducibility, and long-term preservation [29]. These practices have fostered a durable and collaborative research ecosystem—particularly for common malignancies—in which validated cell lines are readily available to the global scientific community.

In contrast, the landscape for rare cancers remains underdeveloped. Despite the recent rise of patient-derived cancer models such as organoids and patient-derived xenografts (PDXs), conventional cancer cell lines continue to serve as the most widely disseminated and experimentally tractable in vitro platforms. However, for many rare tumor subtypes—including sarcomas, pediatric malignancies, and ultra-rare carcinomas—only a small subset of histologies is represented by cell lines that are both described in the scientific literature and deposited in publicly accessible biobanks. This gap presents a major bottleneck for mechanistic investigation and preclinical evaluation.

An effort to quantify this limitation was conducted by Hattori et al., who systematically analyzed the Cellosaurus database—a comprehensive international repository of cell line metadata [30]. Their study revealed that, among sarcomas, only approximately 100 cell lines met the dual criteria of being both reported in publication and registered in public cell banks [31]. This highlights the disproportionate scarcity of robust model systems available for one of the most histologically diverse families of rare cancers.

A key bottleneck in advancing rare cancer research is the persistent lack of patient-derived cancer models, which has led to a structural negative feedback loop that undermines both scientific progress and translational potential [32]. Without appropriate experimental models, mechanistic studies and therapeutic explorations remain limited. This, in turn, makes it difficult to attract new researchers or secure funding, thereby further delaying the development of new models and reinforcing the cycle. Several systemic challenges underlie this issue. Clinical specimens from rare cancers are inherently difficult to obtain, and model establishment requires specialized technical expertise, infrastructure, and sustained efforts, often with low success rates. However, such foundational activities are frequently underrecognized in academic evaluation systems and underfunded in national and international research programs. To break this cycle, targeted policy interventions are urgently needed. Public research agencies and funding bodies should allocate dedicated resources within cancer research budgets to support systematic biobanking, the establishment of patient-derived models, and shared infrastructure across institutions. Technical training programs and collaborative frameworks can reduce duplication and improve success rates. Moreover, academic institutions and grant agencies should update evaluation criteria to appropriately value model development efforts, recognizing their essential role in enabling future discovery. Including model development as a strategic priority in national cancer control plans and international funding initiatives would provide the structural support needed to sustain this work. Strategic investment in these upstream capacities can catalyze downstream research, attract a new generation of scientists to the field of rare cancers, and ultimately improve outcomes for underserved patient populations.

In the sections that follow, we expand upon this analysis by focusing on osteosarcoma—a prototypical rare malignancy—as a case study. We provide a detailed inventory of the osteosarcoma cell lines that are currently available, examining their publication status, accessibility through public repositories, and extent of molecular and functional characterization. Through this focused evaluation, we aim to illuminate the broader structural challenges and unmet needs that confront rare cancer researchers working to establish and share high-quality, biologically relevant cell line models.

9.4 Osteosarcoma: Clinical Features and Therapeutic Landscape

Osteosarcoma is the most common primary malignant bone tumor, primarily affecting adolescents and young adults [33]. It arises most frequently in the metaphyseal regions of long bones, notably the distal femur, proximal tibia, and proximal humerus. The defining histopathological feature of osteosarcoma is the production of osteoid by malignant mesenchymal cells. Diagnosis is established through a combination of radiographic imaging, histopathological assessment, and immunohistochemical analysis. Clinically, the disease exhibits a bimodal age distribution, with incidence peaks around 18 years of age and again in individuals over 60.

The genomic landscape of osteosarcoma is exceptionally complex, distinguished by pervasive chromosomal instability and structural rearrangements rather than recurrent single-nucleotide variants. This genomic disarray is typified by widespread copy number alterations, including high-level amplifications of oncogenes such as MYC, MDM2, and CDK4, and recurrent deletions affecting key tumor suppressors such as TP53, RB1, and CDKN2A [34, 35]. Among these alterations, disruption of the TP53 pathway emerges as a nearly ubiquitous feature and is widely regarded as a central and early event in osteosarcoma pathogenesis [36]. Additionally, chromothripsis, a catastrophic genomic phenomenon involving extensive chromosome fragmentation and reassembly, is frequently detected and may be instrumental in shaping the nonlinear evolutionary trajectory of osteosarcoma [37]. These features collectively suggest a pathogenic mechanism driven not by sequential mutation accumulation but by episodic genomic crises and defective DNA repair.

Prior to the mid-1970s, amputation was the standard treatment for localized disease, and 5-year survival rates were below 20%. The advent of adjuvant and neoadjuvant chemotherapy in the 1970s marked a major therapeutic breakthrough, increasing survival rates to approximately 50%. By the 1990s, the integration of multi-agent chemotherapy with limb-salvage surgery had become standard practice, further improving a 5-year survival to over 65% in patients with localized disease [34, 35]. To date, the approval of novel anticancer agents for osteosarcoma has been exceedingly rare, reflecting both the biological complexity of the disease and the inherent challenges of conducting clinical trials in rare cancers. Although several investigational therapies—including antibody–drug conjugates, immune-modulatory agents, and molecularly targeted drugs—are currently under development and show early signs of efficacy, none have yet reshaped the standard of care. Widespread clinical implementation of these agents will require continued investment in translational research, well-designed multicenter trials, and concerted efforts to navigate the regulatory landscape. This stagnation underscores the urgent need for innovative approaches that can accelerate therapeutic discovery in osteosarcoma, including the integration of preclinical models, multi-omics profiling, and collaborative research frameworks tailored to the unique challenges of rare malignancies.

9.5 Landscape and Accessibility of Osteosarcoma Cell Lines

To elucidate the broader challenges associated with rare cancer models, we conducted a focused analysis of osteosarcoma cell lines—a representative example of a rare yet clinically significant malignancy. Specifically, we sought to quantify the number of osteosarcoma cell lines established to date, assess their documentation in the scientific literature, and evaluate their availability through public biorepositories. For this purpose, we extracted and systematically analyzed data from the Cellosaurus database (version 50, September 2024). The analytic workflow is summarized in Fig. 9.1.

As of September 2024, Cellosaurus documents 159,463 cell lines in total, including 118,612 of human origin, of which 85,865 are categorized as original unique lines. Among these, 226 entries are annotated as osteosarcoma-derived. However, closer inspection reveals notable shortcomings in both traceability and accessibility. Only 176 of these lines (77.8%) have been cited in the publications, leaving 50 lines (22.1%) without formal documentation in the publication. Moreover, although 69 osteosarcoma lines are listed in public cell banks, only 27 are currently verifiable as available for distribution (39.1%). A total of 42 lines were once banked and discontinued. These observations suggested the fragility of laboratory-generated resources and the challenges of ensuring their long-term accessibility.

The discrepancy between the number of established osteosarcoma cell lines and the smaller subset that are both published and distributable constitutes a substantial bottleneck in model reproducibility and translational applicability. This issue is particularly acute for rare cancers, where establishing even a single model requires significant clinical coordination and biospecimen access. The failure to archive and disseminate these models represents a loss of irreplaceable biological material and an impediment to scientific advancement. Addressing this gap will require deliberate and coordinated efforts to register, characterize, and deposit osteosarcoma cell lines into publicly accessible repositories, thereby enhancing their utility across the continuum of cancer research.

A more granular analysis of publication and deposition trends underscores the fragmented nature of the current landscape. Of the 226 osteosarcoma lines catalogued, those associated with publications outnumber unpublished lines by more than threefold (176 vs. 50), reaffirming the central—though incomplete—role of the publication in model dissemination. Yet paradoxically, the proportion of published lines that are deposited in cell banks (30, 17.0%) is only marginally lower than that of unpublished lines (39, 78.0%), suggesting that publication status alone is not a

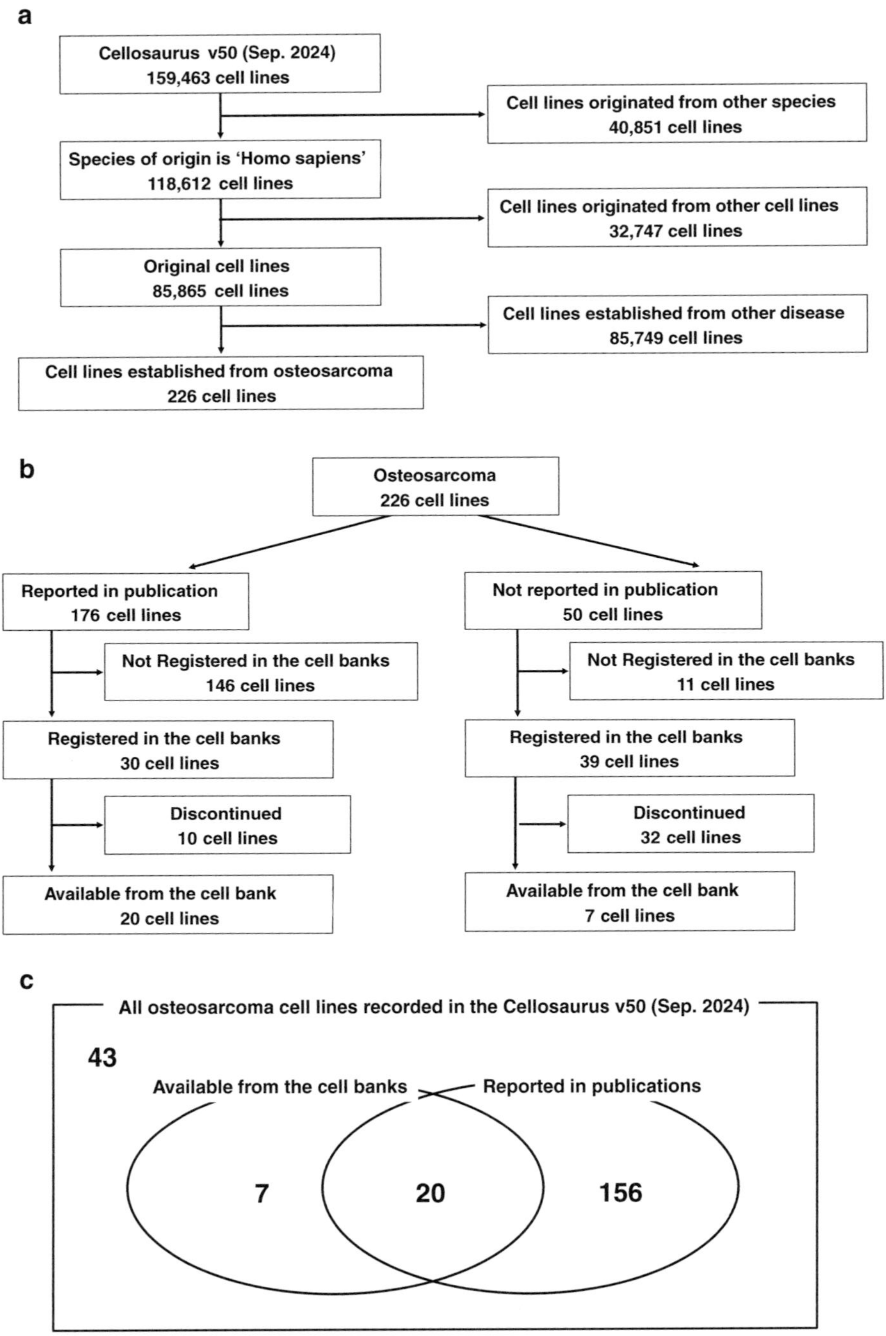

Fig. 9.1 Stratified analysis of osteosarcoma cell lines recorded in Cellosaurus (version 50, September 2024). (**a**) Classification of all cell lines and osteosarcoma-specific entries. Cellosaurus version 50 (September 2024) contains 159,463 cell lines in total, including 118,612 derived from

reliable predictor of availability. This disconnect may reflect temporal lags, institutional dissemination policies, or inconsistent prioritization of long-term resource sharing.

Notably, unpublished lines appear particularly vulnerable to discontinuation. Of the 39 cell lines without publications, 32 (82.1%) have already been retired, compared to 10 of the 30 published lines (33.3%). While the reasons for attrition are varied—including quality control failures, redundancy, or limited utility—these data suggest that cell lines lacking publication are less likely to be preserved. The absence of a published record may limit both scientific credibility and perceived value, contributing to their neglect.

Even among published models, public accessibility remains strikingly limited. Only 20 of the 176 published lines (11.4%) and just 7 of the 50 unpublished lines (14.0%) are currently confirmed as available. In total, only 27 of 226 cell lines (11.9%) are verifiably accessible—an alarmingly low figure that highlights systemic inefficiencies in resource stewardship. Addressing this deficit will require not only improvements in archiving and authentication but also cultural and structural shifts to promote sharing and long-term preservation.

9.6 Limitations of Available Models

Given the pronounced clinical, histopathological, and molecular heterogeneity of osteosarcoma, the limited number of publicly available cell lines—currently only 27—represents a critical bottleneck in translational research. This scarcity stands in stark contrast to the scale required for clinical evaluation, where therapeutic efficacy must typically be demonstrated across several hundred patients to account for intertumoral variability. To approximate such diversity preclinically, a proportionate expansion in disease-representative models is imperative. Although molecular profiling of existing cell lines can be undertaken by downstream users, essential clinical and histological metadata—such as the patient's age, tumor location, treatment history, and precise histopathological subtype—reside with the originating investigators and are seldom disclosed in sufficient detail. In light of the urgent need for subtype-specific therapeutic strategies in osteosarcoma, the rigorous documentation of clinical context and histological classification should be regarded as an integral

Fig. 9.1 (continued) *Homo sapiens* and 85,865 considered original (non-derivative). Of these, 226 cell lines were established from osteosarcoma specimens. (**b**) Publication, registration, and availability status of osteosarcoma cell lines. Among the 226 osteosarcoma cell lines, 176 have been reported in publications and 50 remain unpublished. A total of 69 lines are registered in public cell banks, 42 were discontinued, and 27 are currently available for distribution. (**c**) Comparative breakdown of publication status and bank availability. This panel shows the overlap between publication and availability from public cell banks. Of the 176 published lines, 20 are available from cell banks. Among the 50 unpublished lines, 43 are unavailable and only 7 are distributed from public cell banks. The figure highlights the limited number of osteosarcoma cell lines that are ultimately accessible to the research community

requirement in the formal publication of newly established cell lines. Only through comprehensive and standardized reporting can these models fulfil their intended role in recapitulating disease complexity and informing precision oncology.

To further characterize the scope and limitations of existing resources, we conducted a detailed review of the 27 osteosarcoma cell lines currently available from public cell banks (Table 9.1). These models, while foundational, reveal demographic and anatomical biases that constrain their generalizability.

Demographically, a greater number of cell lines were derived from male ($n = 16$) than female ($n = 11$) donors—a distribution that may reflect the male predominance of osteosarcoma. Age-wise, the majority of models were established from pediatric or adolescent and young adult (AYA) patients. Only a single line (Hs 890.T) was derived from an older adult. This underrepresentation of elderly patients is particularly concerning given the distinct molecular alterations and clinical trajectories observed in this subgroup, whose tumors are often characterized by complex karyotypes and poor treatment responsiveness.

Anatomically, the majority of cell lines ($n = 22$) originated from primary tumors, while only five were derived from metastatic sites. Among these, just two—Hs 899(C).T and Hs 899(D).T—were obtained from pulmonary metastases, despite the lung being the most frequent and clinically significant site of osteosarcoma dissemination. The dearth of metastatic models presents a critical barrier to studying advanced disease, resistance mechanisms, and progression biology.

These findings highlight a disconnect between the clinical heterogeneity of osteosarcoma and the currently available in vitro platforms. While some biases may reflect epidemiological patterns, the skewed representation of donor age and tumor stage limits the relevance of these models to real-world disease diversity. Bridging this gap requires intentional efforts to generate cell lines from underrepresented patient populations—especially elderly individuals—and from metastatic lesions. Doing so will expand the translational utility of osteosarcoma models and enable more nuanced, biomarker-guided therapeutic development.

Bridging this gap will require deliberate and coordinated efforts to establish new models from underrepresented populations and tumor sites. Priority should be given to the derivation and public dissemination of cell lines from older patients and from metastatic lesions—especially pulmonary metastases—to ensure that preclinical platforms more accurately reflect the real-world heterogeneity of osteosarcoma. Such diversification is essential to advancing biomarker-guided drug development and improving outcomes across the full continuum of disease.

In conducting this study, we relied extensively on the Cellosaurus database, a manually curated and widely utilized repository of cell line metadata. However, we found that the references listed for individual cell lines often corresponded to downstream studies employing the models, rather than the original reports of their derivation. To accurately determine the provenance of each cell line, we performed systematic PubMed searches to identify primary establishment publications. Upon closer examination of 20 osteosarcoma cell lines for which Cellosaurus listed reference articles, we found that no original establishment publication could be identified for six lines (SaOS-2, Hs 704.T, Hs 707(A).T, Hs 88.T, SJSA-1, and NOS-10)

Table 9.1 Osteosarcoma cell lines deposited in public biobanks and their publication status in Cellosaurus

Category	Cell line name	Accession	Cell bank ID	Gender	Age	Derived site	References
A	CHA59	CVCL_GZ61	NCI-DTP; CHA-59	Male	16	Primary; bone	[38]
	U2OS	CVCL_0042	Abcam; ab263976 Abcam; ab279979 ABM; T8293 ATCC; HTB-96 BCRC; 60187 BCRJ; 0304 CLS; 300364 DSMZ; ACC-785 ECACC; 92022711 ICLC; HTL11001 KCB; KCB 200962YJ KCLB; 30096 Ubigene; YC-C030	Female	15	Primary; bone, tibia	[39]
	HOS	CVCL_0312	ATCC; CRL-1543 BCRC; 60308 BCRJ; 0339 CCTCC; GDC0251 CCTCC; GDC0333 CLS; 300449 ECACC; 87070202 ICLC; HTL04003 IZSLER; BS TCL 69 JCRB; IFO50106 KCB; KCB 2015015YJ RCB; RCB0992	Female	13	Primary; bone, right femur	[40]

(continued)

Table 9.1 (continued)

Category	Cell line name	Accession	Cell bank ID	Gender	Age	Derived site	References
	MG-63	CVCL_0426	AddexBio; C0004007/4991 ATCC; CRL-1427 BCRC; 60279 BCRJ; 0173 CCTCC; GDC0074 CLS; 300441 ECACC; 86051601 ICLC; HTL99003 IZSLER; BS TCL 40 JCRB; IFO50108 KCB; KCB 200528YJ KCLB; 21427 NCBI_Iran; C555 RCB; RCB1890 TKG; TKG 0294	Male	14	Metastatic; bone, left femur	[41]
	SaOS-2	CVCL_0548	ATCC; HTB-85 BCRJ; 0217 CancerTools; 161931 CCTCC; GDC0075 CLS; 300331 DSMZ; ACC-243 ECACC; 89050205 ICLC; HTL01001 IZSLER; BS TCL 90 KCB; KCB 200550YJ KCLB; 30085 NCBI_Iran; C453 RCB; RCB0428 RCB; RCB3688 TKG; TKG 0469 Ubigene; YC-D017	Female	11	Primary; bone	Not identified

	Hs 704.T	CVCL_0861	ATCC; CRL-7444	Female	13	Primary; bone	Not identified
	Hs 707(A).T	CVCL_0864	ATCC; CRL-7448	Male	13	Primary; bone	Not identified
	Hs 755(B).T	CVCL_0892	ATCC; CRL-7489_FL	Female	18	Metastatic; bone, proximal tibia	[42]
	Hs 88.T	CVCL_0973	ATCC; CRL-7060	Female	25	Primary; bone	Not identified
	CAL-72	CVCL_1113	DSMZ; ACC-439	Male	10	Primary; bone, patella	[43]
	HuO-3 N1	CVCL_1297	JCRB; JCRB0413 RCB; RCB2104 TKG; TKG 0452	Female	15	Primary; bone	[44]
	HuO9	CVCL_1298	JCRB; JCRB0427	Female	13	Primary; bone, right distal femur	[45]
	NOS-1	CVCL_1610	RCB; RCB1032	Male	16	Primary; bone, left proximal tibia	[46]
	NY	CVCL_1613	JCRB; JCRB0614	Male	15	Primary; bone, right femur	[47]
	SJSA-1	CVCL_1697	AddexBio; C0004004/64 ATCC; CRL-2098 CCTCC; GDC0257 CLS; 305096	Male	19	Primary; bone, femur	Not identified
	G-292 clone A141B1	CVCL_2909	ATCC; CRL-1423 BCRC; 60225 ECACC; 90110522 JCRB; IFO50107 KCLB; 21423 NCBI_Iran; C116	Female	9	Primary; bone	[48]
	HuO9N2	CVCL_2958	JCRB; JCRB0428 RCB; RCB2532 TKG; TKG 0451	Female	13	Primary; bone, right distal femur	[45]
	NOS-2	CVCL_3351	RCB; RCB1033	Male	11	Primary; bone, right distal femur	[46]
	NOS-10	CVCL_4663	RCB; RCB2348	Male	15	Primary; bone	Not identified
	HS-Os-1	CVCL_8716	RCB; RCB2229	Female	11	Metastatic; bone, left humerus	[49]

(continued)

Table 9.1 (continued)

Category	Cell line name	Accession	Cell bank ID	Gender	Age	Derived site	References
B	CCLF-PEDS-0019-T	CVCL_E1HD	DSMZ; ACC-970	Male	16	Primary; bone, femur	Not identified
	Hs 184.T	CVCL_0668	ATCC; CRL-7134	Male	21	Primary; bone	Not identified
	Hs 39.T	CVCL_0740	ATCC; CRL-7023	Male	17	Primary, bone	Not identified
	Hs 890.T	CVCL_0984	ATCC; CRL-7628	Male	40	Primary; bone	Not identified
	Hs 894(A).T	CVCL_0987	ATCC; CRL-7631_FL	Male	20	Primary; bone	Not identified
	Hs 899(C).T	CVCL_0998	ATCC; CRL-7644_FL	Male	20	Metastatic; lung	Not identified
	Hs 899(D).T	CVCL_0999	ATCC; CRL-7645_FL	Male	20	Metastatic; lung	Not identified
This table lists 27 osteosarcoma cell lines currently deposited in public biobanks							
Category A comprises 20 cell lines for which at least one publication is registered in Cellosaurus, while Category B includes 7 cell lines with no publication currently listed							
The establishment of these cell lines is documented in the references cited							

This table lists 27 osteosarcoma cell lines currently deposited in public biobanks. Category A comprises 20 cell lines for which at least one publication is registered in Cellosaurus, while Category B includes 7 cell lines with no publication currently listed. The establishment of these cell lines is documented in the references cited and that only a meeting abstract appeared to exist for one line (G-292 clone A141B1). In many other instances, detailed molecular and phenotypic characterizations were reported not in the initial derivation papers but in subsequent functional studies. For cell lines lacking any cited references in Cellosaurus, even exhaustive manual searches failed to yield relevant documentation, underscoring persistent gaps in the formal reporting and archival of cell line establishment. It highlights the pressing need for standardized reporting practices and more efficient editorial requirements to ensure that newly established cell lines are appropriately documented, and their presence is shared in the research community.

9.7 The Critical Importance and Practical Considerations of Publishing Newly Established Cell Lines in Peer-Reviewed Literature

The publication of newly established cancer cell lines is not a ceremonial academic exercise, but a scientific imperative. It underpins transparency and long-term utility—principles that are particularly vital in the context of rare cancers, where each experimental model represents a scarce and valuable resource. Although thousands of cancer cell lines have been generated worldwide, a substantial proportion remain unpublished, uncharacterized, and inaccessible. This lack of formal dissemination constitutes a major bottleneck in the scientific utility of these models, constraining their contribution to basic, translational, and clinical research.

The following sections outline the principal reasons why formal publication remains indispensable for maximizing the scientific value and translational potential of newly established cell lines.

First, publication serves as a critical mechanism for scientific validation. It provides a transparent framework through which the provenance, credibility, and biological relevance of new models can be rigorously assessed. Standard publications typically include essential metadata—such as donor clinical characteristics, histopathological confirmation, and authentication protocols including STR profiling and mycoplasma testing—thereby enabling the broader research community to evaluate the utility of the resource.

Second, publication enhances the discoverability and accessibility of cell lines. Models described in publications are more likely to be indexed in global registries such as Cellosaurus and more frequently deposited in public biorepositories, thus facilitating their integration into collaborative research efforts, including initiatives such as the Cancer Cell Line Encyclopedia (CCLE) and the Cancer Dependency Map (DepMap). Conversely, unreported lines often remain isolated within individual laboratories, rarely subjected to external validation and largely excluded from the broader biomedical research ecosystem. Furthermore, publication confers scholarly recognition to the investigators and institutions responsible for generating the models, reinforcing incentive structures that encourage data sharing and open resource distribution.

Third, publication establishes an ethical and procedural framework for the responsible use of patient-derived biological materials. Reports typically detail protocols for ethical oversight, including informed consent and institutional review board (IRB) approval, and promote methodological transparency. In doing so, they align with increasing international expectations for accountability in the use of human biospecimens.

Fourth, publication contributes to the long-term sustainability of cell line resources. Models that are formally documented in the literature are more likely to be maintained, deposited, and incorporated into curated biobanking infrastructures. By contrast, unpublished models face a disproportionately high risk of attrition due to poor documentation, lack of visibility, and minimal institutional support. This pattern is exemplified in osteosarcoma, where unpublished cell lines exhibit markedly elevated rates of discontinuation compared to their published counterparts (Fig. 9.1b)—highlighting the long-term consequences of remaining outside the formal scientific record.

Nevertheless, while publication is essential, it must also be made feasible. Excessive editorial or reviewer demands for comprehensive molecular characterization—particularly at the time of initial reporting—may paradoxically suppress dissemination. It is important to distinguish between foundational data necessary to establish the identity and provenance of a cell line and application-specific datasets that can and should emerge through subsequent research. For example, while omics technologies such as genomics, transcriptomics, and proteomics provide valuable biological insights, expecting such data at the time of model establishment conflates primary validation with downstream functional analysis. Moreover, omics platforms and analytic standards evolve rapidly; thus, early-generation data may quickly become obsolete or difficult to reproduce. In contrast, core information—such as donor demographics, tissue origin, method of isolation, STR profiling results, and mycoplasma testing status—retains enduring relevance and ensures model traceability across research contexts and timeframes.

The goal of a cell line establishment paper is not to exhaustively explore all potential applications but to provide minimally sufficient documentation to make the model accessible, reliable, and interoperable. Historical precedents affirm that many of today's most widely used cell lines were initially published with limited characterization and that further profiling accumulated progressively through community-driven use and study. Accordingly, editors and reviewers should avoid imposing requirements for niche or technically demanding assays that are only feasible in select laboratories, or for experiments unlikely to be conducted in the near term. Setting the bar for publication unreasonably high may disincentivize formal reporting altogether—ironically exacerbating the very opacity and fragmentation that rigorous publication is intended to overcome.

There is a clear need for practical guidelines that define the essential data sufficient for publishing newly established cancer cell lines. By articulating such a "minimum standard," researchers can avoid excessive publication demands while ensuring transparency, reproducibility, and accessibility of these valuable resources.

This is particularly crucial in rare cancers, where the number of cases and available models is inherently limited. Table 9.2 presents a set of data items that, if fully provided, should be considered sufficient for publication. While additional data—such as in vivo tumorigenicity or omics profiling—may enhance a manuscript, they should not be regarded as prerequisites when the core elements listed in Table 9.2 are adequately documented. Requiring data that only a small subset of researchers

Table 9.2 Sufficient data requirements for publishing newly established cancer cell lines

Category	Required data	Notes
Clinical and specimen origin	– Patient gender and age – Tumor type and anatomical location – Clinical history (genetic backgrounds if applicable, treatment details) – Sample acquisition method and timing – Statement of informed consent and IRB approval	Ensures traceability and ethical compliance. The clinical and pathological data required for translational research may evolve over time. Authors should strive to provide the most comprehensive and precise information available at the time of publication, with consideration for potential future research needs
Histological verification	– H&E staining of parental tumor – Diagnosis confirmed by pathologist – Optional immunohistochemistry	Establishes lineage and authenticity. To ensure future re-evaluation in light of evolving diagnostic criteria, inclusion of raw histological images in the publication is strongly recommended
Culture characteristics	– Morphological description (e.g., adherent vs. floating) – Growth characteristics (e.g., doubling time) – Culture conditions (medium, additives, environment)	Facilitates reproducibility and model suitability. While cell lines can be used in a vast array of downstream experiments, demonstrating such applications is not the purpose of a cell line establishment paper. The focus should remain on the foundational attributes of the model itself
Authentication	– STR profiling with database comparison – Mycoplasma testing (must be negative)	STR profiles of both the cell line and its parental tumor should be included to verify derivation and prevent misidentification. Mycoplasma-free status must be demonstrated prior to publication
Availability and registration	– Unique cell line name – Repository registration (if available) – Cellosaurus entry (if available) – Conditions for distribution (e.g., MTA)	While the inclusion of availability information is encouraged, it is understood that institutional or legal policies may restrict open distribution. Therefore, disclosure of accessibility is recommended but not mandatory
Optional but recommended data	– Genetic features (e.g., mutations, fusion genes) – Drug sensitivity data (optional screening results) – Omics profiling data with database deposition information	Encouraged to support downstream research and cross-study validation. However, as omics data types may become obsolete with evolving technologies, these elements are not considered mandatory for publication

use, or insisting on experiments that may fall out of standard practice as methodologies evolve, is both inefficient and detrimental to scientific progress. Such practices risk raising unnecessary barriers and discouraging the dissemination of valuable research tools. It is essential that the publication of new cell lines be guided by consistent, realistic, and forward-compatible standards that reflect the foundational requirements for usability and validation—nothing more and nothing less. The author has, in fact, encountered requests for such nonessential data during the peer review process. If such practices are widespread, they may constitute a systemic obstacle to the establishment and distribution of cell lines for rare cancers, where available resources and opportunities for validation are inherently limited. Recognizing and correcting these inefficiencies is critical to accelerating model development and ensuring equitable progress in cancer research.

There may be differing views on the publication process and the role of biobanks, depending on their intended purpose. However, when it comes to the establishment of new cell lines, stakeholders must recognize that the primary goal is the permanent and shared availability of these valuable resources. In rare cancers such as osteosarcoma, the scarcity of cell lines is a critical issue, and expanding their supply is an urgent priority. Researchers who establish cell lines for rare cancers should view it not as a private asset for personal use but as a shared resource with an obligation to distribute widely for the benefit of the broader scientific community. Ultimately, the goal is not merely to generate cell lines but to embed them within a sustainable, transparent, and equitable research ecosystem. Achieving this vision requires structural reform, cultural shifts, and renewed institutional commitment—particularly in the field of rare cancers, where each model holds disproportionate scientific and therapeutic value.

9.8 Conclusions

In the field of rare cancer research, access to clinical specimens is both a privilege and a responsibility. Given the scarcity of such biospecimens and the urgent need for disease-relevant models, investigators with access to rare cancer tissues arguably have an ethical obligation to prioritize the establishment of cell lines. These models not only serve as critical tools for basic and translational studies but also enable the broader research community to investigate diseases that are otherwise underrepresented in global oncology efforts. Without systematic efforts to convert precious clinical material into shareable experimental resources, the translational potential of rare cancer research remains severely constrained.

To realize this potential, structural reforms are needed in how newly established cell lines are evaluated and disseminated. Excessive demands for comprehensive experimental data at the time of initial publication—while well-intentioned—can create unrealistic hurdles that deter researchers from reporting valuable models. Beyond this, there is an urgent need for clear editorial guidelines to prevent reviewers from requesting experiments that fall outside the scope of identity verification, or that reflect personal interests rather than community standards. Review processes must distinguish between foundational validation—such as donor metadata, STR authentication, and mycoplasma testing—and exploratory data that may develop in subsequent work.

Finally, to promote long-term accessibility, academic and funding institutions should implement incentive structures that reward deposition in public cell banks. Without recognition and support, even well-characterized cell lines risk the remaining invisible and unused. By lowering barriers to publication and aligning incentives toward model sharing, the research community can build a more inclusive and sustainable foundation for rare cancer discovery.

References

1. Working Group on the Provision and Support of Medical Care for Rare Cancers, Report of the Working Group on the Provision and Support of Medical Care for Rare Cancers (in Japanese), 2015.
2. Chia PL, Mitchell P, Dobrovic A, John T. Prevalence and natural history of ALK positive non-small-cell lung cancer and the clinical impact of targeted therapy with ALK inhibitors. Clin Epidemiol. 2014;6:423–32.
3. Cardarella S, Ogino A, Nishino M, Butaney M, Shen J, Lydon C, Yeap BY, Sholl LM, Johnson BE, Jänne PA. Clinical, pathologic, and biologic features associated with BRAF mutations in non-small cell lung cancer. Clin Cancer Res. 2013;19(16):4532–40.
4. RARECANCERNet. http://rarecarenet.istitutotumori.mi.it/rarecarenet/.
5. Rare and less common cancers. https://www.cancer.nsw.gov.au/about-cancer/types-of-cancer/rare-and-less-common-cancers?utm_source=chatgpt.com.
6. DeSantis CE, Kramer JL, Jemal A. The burden of rare cancers in the United States. CA Cancer J Clin. 2017;67(4):261–72.
7. Matsuda T, Won YJ, Chun-Ju Chiang R, Lim J, Saika K, Fukui K, Lee WC, Botta L, Bernasconi A, Trama A. Rare cancers are not rare in Asia as well: the rare cancer burden in East Asia. Cancer Epidemiol. 2020;67:101702.
8. Gatta G, van der Zwan JM, Casali PG, Siesling S, Dei Tos AP, Kunkler I, Otter R, Licitra L, Mallone S, Tavilla A, et al. Rare cancers are not so rare: the rare cancer burden in Europe. Eur J Cancer. 2011;47(17):2493–511.
9. Keat N, Law K, Seymour M, Welch J, Trimble T, Lascombe D, Negrouk A. International rare cancers initiative. Lancet Oncol. 2013;14(2):109–10.
10. Panageas KS. Clinical trial design for rare cancers: why a less conventional route may be required. Expert Rev Clin Pharmacol. 2015;8(6):661–3.
11. Roman YM. Pharmacogenomics and rare diseases: optimizing drug development and personalized therapeutics. Pharmacogenomics. 2025;26(3–4):121–8.
12. Gatta G, Capocaccia R, Botta L, Mallone S, De Angelis R, Ardanaz E, Comber H, Dimitrova N, Leinonen MK, Siesling S, et al. Burden and centralised treatment in Europe of rare tumours: results of RARECAREnet—a population-based study. Lancet Oncol. 2017;18(8):1022–39.
13. de Heus E, Duijts SFA, van der Zwan JM, Kapiteijn E, Nieveen van Dijkum EJM, van Herpen CML, Merkx MAW. The gap between rare and common cancers still exists: results from a population-based study in the Netherlands. Eur J Cancer. 2022;167:103–11.
14. Low CE, Loke S, Pang GE, Sim B, Yang VS. Psychological outcomes in patients with rare cancers: a systematic review and meta-analysis. EClinicalMedicine. 2024;72:102631.
15. Rare Cancer Center. https://www.ncc.go.jp/en/ncch/division/rcc/index.html.

16. Vivelo C, Reilly KM, Widemann BC, Wedekind MF, Painter C, O'Neill AF, Mueller S, Elemento O, Gross AM, Sandler AB. The landscape of US and global rare tumor research programs: a systematic review. Oncologist. 2024;29(2):106–16.
17. The MASTER KEY PROJECT Genomic medicine in rare cancers: a collaboration between the industry and academia. https://www.ncc.go.jp/en/information/press_release/2017/0731/index.html.
18. Heo YJ, Hwa C, Lee GH, Park JM, An JY. Integrative multi-omics approaches in cancer research: from biological networks to clinical subtypes. Mol Cells. 2021;44(7):433–43.
19. Box GEP. Science and statistics. J Am Stat Assoc. 1976;71(356):791–9.
20. Jin H, Zhang C, Zwahlen M, von Feilitzen K, Karlsson M, Shi M, Yuan M, Song X, Li X, Yang H, et al. Systematic transcriptional analysis of human cell lines for gene expression landscape and tumor representation. Nat Commun. 2023;14(1):5417.
21. Cancer Cell Line Encyclopedia (CCLE). https://sites.broadinstitute.org/ccle.
22. Ghandi M, Huang FW, Jané-Valbuena J, Kryukov GV, Lo CC, McDonald ER III, Barretina J, Gelfand ET, Bielski CM, Li H, et al. Next-generation characterization of the Cancer Cell Line Encyclopedia. Nature. 2019;569(7757):503–8.
23. Barretina J, Caponigro G, Stransky N, Venkatesan K, Margolin AA, Kim S, Wilson CJ, Lehár J, Kryukov GV, Sonkin D, et al. The Cancer Cell Line Encyclopedia enables predictive modelling of anticancer drug sensitivity. Nature. 2012;483(7391):603–7.
24. Genomics of Drug Sensitivity in Cancer (GDSC). https://www.cancerrxgene.org/.
25. Tsherniak A, Vazquez F, Montgomery PG, Weir BA, Kryukov G, Cowley GS, Gill S, Harrington WF, Pantel S, Krill-Burger JM, et al. Defining a Cancer Dependency Map. Cell. 2017;170(3):564–576.e516.
26. Arafeh R, Shibue T, Dempster JM, Hahn WC, Vazquez F. The present and future of the Cancer Dependency Map. Nat Rev Cancer. 2025;25(1):59–73.
27. Le NN, Frater I, Lip S, Padmanabhan S. Hypertension precision medicine: the promise and pitfalls of pharmacogenomics. Pharmacogenomics. 2025;26(5–6):183–206.
28. Savage SR, Yi X, Lei JT, Wen B, Zhao H, Liao Y, Jaehnig EJ, Somes LK, Shafer PW, Lee TD, et al. Pan-cancer proteogenomics expands the landscape of therapeutic targets. Cell. 2024;187(16):4389–4407.e4315.
29. Weiskirchen S, Monteiro AM, Borojevic R, Weiskirchen R. Unlocking potential: a comprehensive overview of cell culture banks and their impact on biomedical research. Cells. 2024;13(22):1861.
30. Bairoch A. The Cellosaurus, a cell-line knowledge resource. J Biomol Tech. 2018;29(2):25–38.
31. Hattori E, Oyama R, Kondo T. Systematic review of the current status of human sarcoma cell lines. Cells. 2019;8(2):157.
32. Kondo T. Current status and future outlook for patient-derived cancer models from a rare cancer research perspective. Cancer Sci. 2021;112(3):953–61.
33. Kansara M, Teng MW, Smyth MJ, Thomas DM. Translational biology of osteosarcoma. Nat Rev Cancer. 2014;14(11):722–35.
34. Isakoff MS, Bielack SS, Meltzer P, Gorlick R. Osteosarcoma: current treatment and a collaborative pathway to success. J Clin Oncol. 2015;33(27):3029–35.
35. Beird HC, Bielack SS, Flanagan AM, Gill J, Heymann D, Janeway KA, Livingston JA, Roberts RD, Strauss SJ, Gorlick R. Osteosarcoma. Nat Rev Dis Primers. 2022;8(1):77.
36. Saba KH, Difilippo V, Kovac M, Cornmark L, Magnusson L, Nilsson J, van den Bos H, Spierings DC, Bidgoli M, Jonson T, et al. Disruption of the TP53 locus in osteosarcoma leads to TP53 promoter gene fusions and restoration of parts of the TP53 signalling pathway. J Pathol. 2024;262(2):147–60.
37. Espejo Valle-Inclan J, De Noon S, Trevers K, Elrick H, van Belzen I, Zumalave S, Sauer CM, Tanguy M, Butters T, Muyas F, et al. Ongoing chromothripsis underpins osteosarcoma genome complexity and clonal evolution. Cell. 2025;188(2):352–370.e322.

38. Saini V, Hose CD, Monks A, Nagashima K, Han B, Newton DL, Millione A, Shah J, Hollingshead MG, Hite KM, et al. Identification of CBX3 and ABCA5 as putative biomarkers for tumor stem cells in osteosarcoma. PLoS One. 2012;7(8):e41401.
39. Pontén J, Saksela E. Two established in vitro cell lines from human mesenchymal tumours. Int J Cancer. 1967;2(5):434–47.
40. McAllister RM, Gardner MB, Greene AE, Bradt C, Nichols WW, Landing BH. Cultivation in vitro of cells derived from a human osteosarcoma. Cancer. 1971;27(2):397–402.
41. Billiau A, Edy VG, Heremans H, Van Damme J, Desmyter J, Georgiades JA, De Somer P. Human interferon: mass production in a newly established cell line, MG-63. Antimicrob Agents Chemother. 1977;12(1):11–5.
42. Smith HS, Owens RB, Hiller AJ, Nelson-Rees WA, Johnston JO. The biology of human cells in tissue culture. I. Characterization of cells derived from osteogenic sarcomas. Int J Cancer. 1976;17(2):219–34.
43. Rochet N, Dubousset J, Mazeau C, Zanghellini E, Farges MF, de Novion HS, Chompret A, Delpech B, Cattan N, Frenay M, et al. Establishment, characterisation and partial cytokine expression profile of a new human osteosarcoma cell line (CAL 72). Int J Cancer. 1999;82(2):282–5.
44. Yamane T. Establishment and characterization of cell lines derived from a human osteosarcoma. Clin Orthop Relat Res. 1985;199:261–71.
45. Kawai A, Ozaki T, Ikeda S, Oda T, Miyazaki M, Sato J, Taketa K, Inoue H. Two distinct cell lines derived from a human osteosarcoma. J Cancer Res Clin Oncol. 1989;115(6):531–6.
46. Hotta T, Motoyama T, Watanabe H. Three human osteosarcoma cell lines exhibiting different phenotypic expressions. Acta Pathol Jpn. 1992;42(8):595–603.
47. Sekiguchi M, Asanuma K, Satomura T, Fukushima H, Fujji G, Shimoda T, Fukunaga M, Ishikawa E. A cultured cell line established in vitro from human osteosarcoma. Jpn J Exp Med. 1983;53(6):289–92.
48. Peebles PT, Trisch T, Papageorge AG. 727 ISOLATION OF FOUR UNUSUAL PEDIATRIC SOLID TUMOR CELL LINES. Pediatr Res. 1978;12(4):485.
49. Sonobe H, Mizobuchi H, Manabe Y, Furihata M, Iwata J, Hikita T, Kiuna O, Tanimoto T, Oka T, Ohtsuki Y. [Establishment and characterization of a cell line, HS-Os-1 derived from an osteoblastic type of human osteosarcoma]. Hum Cell. 1990;3(2):148–151.

Chapter 10
Applications of Induced Pluripotent Stem Cells in Neurodegenerative Disease Drug Discovery and Development with Artificial Intelligence

Yukari Blumenthal, Keiko Imamura, and Haruhisa Inoue

Abstract The most attributable risk factor for the pathogenesis of neurodegenerative diseases is aging, which is also one of the greatest challenges in a super-aging society. With the development of biological and genetic research in recent years, our understanding of the molecular biology of neurodegenerative diseases has significantly advanced; however, for neurodegenerative diseases, especially for sporadic cases without an identifiable family history, which account for the majority of affected patients, effective models to investigate the pathogenesis have been needed. Patient-specific induced pluripotent stem cells (iPSC) generated by using reprogramming technology have the identical genetic background as that of the patients' and can be differentiated into any cell types to construct a disease cell model. Analysis of the omics data, including cell-phenotype data, may lead to elucidation of the disease pathophysiology and identification of potential therapeutic targets. In addition, because these data are vast and complex, analyses using artificial intelligence (AI) are being pursued. In this review, we describe research using iPSC data and AI in the field of neurodegenerative diseases, including our own research to provide examples.

Y. Blumenthal
Center for iPS Cell Research and Application, Kyoto University (CiRA), Kyoto, Japan

Astellas Pharma Inc., Chuo-Ku, Tokyo, Japan

Kindness United Neuroscience Organization (K.U.N.O.) e.V., Dortmund, Germany

K. Imamura · H. Inoue (✉)
Center for iPS Cell Research and Application, Kyoto University (CiRA), Kyoto, Japan

Medical-Risk Avoidance Based on iPS Cells Team, RIKEN Center for Advanced Intelligence Project (AIP), Kyoto, Japan

iPSC-Based Drug Discovery and Development Team, RIKEN BioResource Research Center (BRC), Kyoto, Japan
e-mail: haruhisa@cira.kyoto-u.ac.jp

N. Shinomiya, Y. Shimada (eds.), *Applications of Cell Culture*, Current Human Cell Research and Applications, https://doi.org/10.1007/978-981-95-0587-6_10

Keywords Induced pluripotent stem cell (iPSC) · Artificial intelligence (AI) · Cellular dissection of polygenicity (CDiP) · Deep learning with iPSC-based scanning (Deep iScan) · Prediction model based on the heat-diffusion equation (PM-HDE)

Abbreviations

AD	Alzheimer's disease
AI	Artificial intelligence
ALS	Amyotrophic lateral sclerosis
ALSFRS	The Amyotrophic Lateral Sclerosis Functional Rating Scale
ALSFRS-R	Revised Amyotrophic Lateral Sclerosis Functional Rating Scale
ADNI	Alzheimer's disease neuroimaging initiative
AxD	Alexander disease
CDiP	Cellular dissection of polygenicity
FTLD	Frontotemporal lobar degeneration
HD	Huntington's disease
HDE	Hierarchical dynamics encoder
iPSCs	induced pluripotent stem cells
LightGBM	Light gradient boosting machine
PD	Parkinson's disease
PET	Positron emission tomography
RF	Random forest
SCP	Single-cell proteomics
SNP	Single nucleotide polymorphism
UMAP	Uniform manifold approximation and projection

10.1 Introduction

Neurodegenerative diseases are intractable diseases in which specific neuronal cells in the brain and spinal cord gradually degenerate, leading to cell degeneration and consequential functional decline. Well-known neurodegenerative diseases include Alzheimer's disease (AD), amyotrophic lateral sclerosis (ALS), and other neurodegenerative diseases in which cognitive and motor functions decline as disease progresses. Pathologically, it is characterized by the death of specific neuronal cells and the accumulation of abnormal proteins. Aging is the strongest and inevitable risk factor for neurodegenerative diseases and is also one of the greatest challenges in a super-aging society. With the development of biological and genetic research in recent years, the molecular biological understanding of neurodegenerative diseases has significantly advanced. However, for sporadic neurodegenerative diseases with no identifiable family history, which account for the majority of affected patients, there have been limited effective models to study the precise causes of pathophysiology.

With the advent of mouse in 2006 and human iPSC in 2007 [1, 2], it has become possible to generate iPSC derived from patients with neurodegenerative diseases and induce differentiation therefrom to better understand the underlying pathophysiology [3–5]. This technology has also been adapted and found highly informative in the plethora of ultra-rare and rare conditions [6, 7].

The prototype of today's deep learning, proposed by Geoffrey Hinton and his colleagues at the University of Toronto in 2006, coincidentally took place at the same time as the birth of mouse iPSC [8]. In 2012, the convolutional neural network model "AlexNet" for object recognition, announced by Hinton's team, won the first place in an image classification competition, which brought attention to deep learning. Since then, AI technologies such as deep learning have brought major changes in the way vast amounts of digital data are processed and analyzed within a society. In the field of disease research, AI is being put to practical use in various aspects such as elucidation of disease causes, diagnosis, prognosis prediction, and drug discovery.

This review introduces neurodegenerative disease research using iPSC and the use of AI in neurodegenerative disease research. Finally, we will discuss research using iPSC data and AI for neurodegenerative diseases, including AD, ALS, and other neurodegenerative diseases.

10.2 Neurodegenerative Disease Research Using iPSC

Due to the difficulty of collecting living brain tissue, it has been a challenge to obtain human neural cells in the form of living cells. However, one of the advantages of research using iPSC is that it is possible to obtain neural cells with the same genetic background as the patient and their phenotypic data from iPSC generated from the patient themselves. To date, various cell models created from iPSC derived from patients with neurodegenerative diseases have been reported.

In AD, the typical pathological findings include the deposition of senile plaques, a histpathological hallmark of AD, mainly composed of amyloid β protein and the abnormal accumulation of phosphorylated tau proteins. Phenotypes such as an increase in the amyloid β42/40 ratio involved in accumulation and increased phosphorylation of tau proteins were reproduced [9–11]. In Parkinson's disease (PD), the main pathological findings are degeneration of dopaminergic neurons in the substantia nigra of the midbrain and the appearance of Lewy bodies whose main component is α-synuclein protein. Abnormal accumulation of α-synuclein protein was also observed in the generated dopaminergic neurons [12, 13]. Similarly, the pathological findings in ALS patients include the accumulation of TDP-43 and loss of motor neurons, reported in iPSC-derived motor neurons [14–16].

In addition, the involvement of glial cells, which are cell types other than neurons, in the pathology of neurodegenerative diseases has been reproduced. Abnormal cytokine secretion, suppressed lactase secretion, and increased oxidative stress were reported in astrocytes derived from iPSC of the patients affected by AD [17]. Among various iPSC models, astrocytes derived from patients affected by PD, exhibit accumulation of α-synuclein protein and abnormal autophagy [18], and oligodendrocytes derived from patients affected by ALS exhibit decreased production and secretion of lactase [19]. The corresponding phenotypes have been reported. In this way, by inducing differentiation of iPSC into each cell type, it is possible to obtain phenotypic data of any cell type produced from iPSC.

Another representative example of this is Alexander disease (AxD). In AxD, primarily astrocytes are being affected; however, holistic assessment of astrocytes in relation to AxD pathophysiology had been challenging due to its extensive and dynamic functions of astrocytes. Aforementioned technological advancement made such analyses of identifying various properties of astrocytes and their complex roles in the disease pathophysiology possible [6, 20] (Table 10.1).

Table 10.1 Summary of applications of iPSC technology

Target identification and validation	Human disease models	iPSCs can be generated from patients with specific diseases, allowing researchers to create in vitro models that closely mimic the disease pathology. For example, iPSCs have been used to model neurodegenerative diseases like AD, ALS, and other neurodegenerative diseases
	Genetic studies	These models enable the study of genetic mutations and their effects on cellular functions, providing insights into disease mechanisms and identifying potential therapeutic targets
	Novel targets	Disease models derived from iPSC can be used to discover and validate new drug targets. This is particularly useful for complex diseases where traditional models have limitations
	Pathway analysis	By studying the pathways involved in disease progression using iPSC models, researchers can identify critical nodes that can be targeted by new drugs
Drug screening and toxicity testing platform	High-throughput/in vitro screening	iPSC can be differentiated into various cell types, such as astrocytes, neurons, or cardiomyocytes, which can be used in high-throughput/in vitro screening assays to test the potential efficacy and safety of new drugs
	Disease-specific assays	iPSC derived from patients with specific genetic backgrounds enable the development of disease-specific screening assays, enhancing the relevance of drug testing
	Personalized medicine	Patient-specific iPSC allow for the testing of drug responses in cells derived from the patient, enabling personalized treatment strategies
Biomarker discovery	Identification of disease biomarkers	iPSC-derived models can help identify biomarkers that may be used for early diagnosis, monitoring disease progression, or objective endpoints in clinical trials
Regenerative medicine and cell therapy	Cell replacement therapy	iPSCs hold potential for regenerating damaged tissues and organs. For example, iPSC-derived dopaminergic neurons are being explored for treating Parkinson's disease
	Gene editing	Combining iPSC with gene editing technologies including CRISPR/Cas9 allows for the correction of genetic defects and the development of gene therapies

10.3 Utilizing Artificial Intelligence in Neurodegenerative Disease Research

In recent years, various omics measurement technologies have developed dramatically, making it possible to quickly obtain more accurate and high-resolution data. At the same time, today we are faced with the need to analyze large amounts of complex data in a shorter period of time, and AI is becoming an indispensable tool. There are various types and methods of AI (Fig. 10.1), and based on the characteristics of each, a method that suits the type of data handled and the purpose of analysis is selected, and highly accurate analysis results are being obtained.

Machine learning is an AI technology in which computers learn from data and make predictions and inferences about new data based on rules and patterns. The deep learning mentioned earlier is a machine learning method for achieving advanced predictions and estimations and uses a multilayer neural network created based on the connections between neurons in the brain. It is particularly effective for advanced pattern recognition of voice and text data.

In the field of neurodegenerative disease research, there have been reports of research using AI to analyze various omics data in order to identify disease-causing genes, support disease diagnosis, and classify patient subtypes.

Stamate et al. reported that it is possible to distinguish between healthy subjects and AD patients by using blood metabolome data or biomarkers such as amyloid β and phosphorylated tau in cerebrospinal fluid as learning data for machine learning [21] . In addition, we are developing a model that uses machine learning to study cognitive function test values and image data of the patients for prognosis; it predicts whether the condition of patients with mild cognitive impairment will progress to AD or remain static as mild cognitive impairment [22]. There is also an example of using machine learning to develop AI that can distinguish between patients

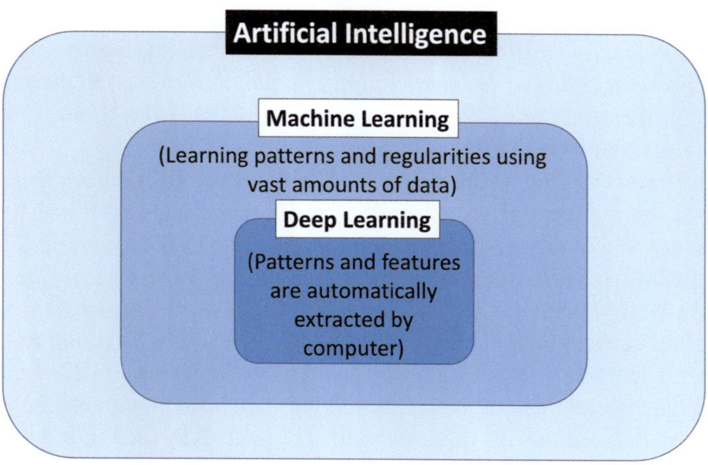

Fig. 10.1 Artificially intelligence classification

affected by PD and healthy individuals based on their breathing patterns during sleep and predict the progression of the disease [23]. Another study was reported by Yin et al. who used deep learning to study whole genome data from healthy individuals and patients affected by ALS and identified promoter regions on chromosomes 7, 9, 17, and 2 that increased the risk of ALS [24]. Zhang et al. obtained transcriptome and epigenome data from motor neurons derived from iPSC of the patients affected by ALS, combined them with genome-wide association analysis data of patients affected by ALS, and performed analysis using machine learning. It identified a novel gene that increased ALS risk [25].

There are also reports that machine learning based on clinical data of patients affected by ALS had been used to classify the patient population into subtypes based on disease progression patterns [26, 27].

By analyzing omics data derived from patients with neurodegenerative diseases using AI, it becomes possible to extract changes that are difficult for humans to discern, and this can lead to various applications of omics data, and its possibilities are expanding with technological advancement.

Furthermore, recently a global concerted effort to develop a novel medical strategy called 37TrillionCells (37TC) based on cell-based interception and precision medicine, in which accelerated detection of neurodegenerative pathogenesis is pursued. In this global consortium, high-performance single-cell proteomics (SCP), pathogenic model cells such as patient derived iPSC models, are combined with predictive computational models using AI for future trajectories, to identify molecular changes well in advance of phenotypical presentation for personalized early therapeutic opitimization [28] (Table 10.2).

10.4 Research Using Neurodegenerative Disease iPS Cell Data and Artificial Intelligence

Pathophysiologically, by the time the symptoms of neurodegenerative diseases become apparent, neuronal degeneration has already progressed to the point of no reversibility; therefore, early therapeutic intervention based on timely diagnosis is paramount in order to seek effective treatments.

We have been working on building a predictive model for AD using patient iPSC data. AD is the resultant of various pathological phenotypes exhibited by various cell types, and one of the cellular phenotypes that deserves attention is the increased amyloid β42/40 ratio in cerebral cortical neurons. We conducted genome-wide association study (GWAS) using changes in the amyloid β42/40 ratio produced from cerebral cortical neurons derived from iPSC (iPSC cohort) comprised of 102 patients with sporadic AD as a pathological phenotype. Using a GWAS, we identified 496 single nucleotide polymorphisms (SNPs) associated with Aβ42/40 in cerebral cortical neurons of the patients with sporadic AD (cell GWAS method). Furthermore, we verified to what extent the iPSC cohort and cell GWAS analysis

Table 10.2 AI-driven neurodegenerative disease studies

Indication	Measured outcome	Parameters	Device/model design	References
AD	To differentiate AD from normal cognition	Plasma metabolites to match the AUC of well-established AD CSF biomarkers	Deep learning (DL), extreme gradient boosting (XGBoost), and random forest classifier	Stamate et al. [21]
AD	To understand pathogenicity of AD, potential therapeutic targets, a rare variant for disease stratification, and polygenes to predict real-word AD	Neuronal production of amyloid β, RWD of AD cohort studies (ADNI and Japanese-ADNI)	CDiP technology, found by a cell GWAS, designed for RWD based predictions via random forest classifier	Kondo et al. [29]
ALS	To identify potential ALS-associated promoter regions	Genotype data—chromosomes 7, 9, 17 and 22	A deep-learning based convolutional neural network architectures—two-step approach to map associations between genotypes and the occurrence of ALS	Yin et al. [24]
ALS	To identify ALS risk genes	Transcriptomic and epigenetic profiling based on GWAS	RefMap identifies ALS risk genes by integrating ALS GWAS data with the molecular profiling of motor neurons	Zhang et al. [25]
ALS	To predict ALS disease progression	ALSFRS slope prediction in PRO-ACT cohort	1. Feed Forward Neural Network (FFNN) 2. Convolutional Neural Network using ALSFS data 3. Recurrent Neural Network	Pancotti et al. [30]

(continued)

Table 10.2 (continued)

Indication	Measured outcome	Parameters	Device/model design	References
ALS	To classify the patterns of functional decline in patients with sporadic ALS and explore the genetic backgrounds that modified these patterns	GWAS of SNPs	A mixture approach of a nonlinear mixed-effects model. Association study between the clusters of longitudinal functional scores and SNPs	Watanabe et al. [26]
ALS	To identify the number and nature of ALS subtypes	ALS Functional Rating Scale (ALSFRS-R) feature values	1. Unsupervised Uniform Manifold Approximation and Projection (UMAP) modeling 2. Semi-supervised UMAP modeling 3. Supervised ensemble learning based on LightGBM model	Faghri et al. [27]
ALS	To support the diagnosis of ALS	Images of iPSC-derived motor neuron	A prediction model of ALS using deep CNN learning algorithms and images of iPSC-derived motor neurons	Imamura et al. [15]
PD	To detect PD and track/predict its progression	Nocturnal breathing signals, sleep, and electroencephalogram	An attention layer that allows for interpreting its predictions	Yang et al. [23]
PD	To identify complex PD-specific disease signatures in human fibroblasts	Fibroblasts from PD patients and matched healthy controls, cell painting image dataset	A platform for the unbiased phenotypic analysis of patient-derived cell lines by applying custom cell culture automation procedures, high-content imaging, and cutting-edge deep learning-based image analysis	Schiff et al. [31]

results reflected actual clinical information (real-world data). The AD Neuroimaging Initiative (ADNI) is a public database of AD that utilizes real-world data (RWD) to precisely evaluate early stages of mild cognitive impairment, which are important targets for prevention and treatment, using image diagnosis and biomarkers. We used Aβ accumulation in the brain and biomarker changes in cerebrospinal fluid registered in ADNI. Among the genetic information of ADNI participants, the 496 SNPs identified by cell GWAS are treated as polygenes and used as input data, and machine learning is used to investigate the accumulation of Aβ in the brain and the brain of actual AD patients. We investigated whether clinical conditions could be predicted, such as changes in biomarkers in the spinal fluid. As a result, compared to using only information on age, sex, and genotype of *APOE*, which has been reported to be involved in Aβ accumulation in the brain, adding information on genotypes of polygenes identified by cell GWAS to these information, it was found to improve the accuracy of disease diagnosis and prediction [32].

In this way, we will reconstruct the disease pathophysiology by obtaining a dataset that links the phenotype of any cell type created from iPSC with the genetic background and analyzing it using machine learning. We proposed Cellular Dissection of Polygenicity (CDiP), a technology enabling this (Fig. 10.2). CdiP technology is expected to make it possible to predict individual disease states [29, 32].

We also constructed a model that can classify ALS using deep learning and image data of iPSC-derived motor neurons [33]. iPSC were established from 15 healthy individuals and 15 patients with sporadic ALS to generate motor neurons. AI built by learning from approximately 7,000 motor neuron image data per person was found to be able to discriminate between motor neuron images of the healthy individuals and that of patients with ALS' with over 90% accuracy (Fig. 10.3). AI recognized motor neurons derived from ALS patients based on cell morphology, particularly the characteristics of cell bodies and neurites [33]. Based on these research results, it is expected that this technology, called Deep learning with iPSC-based scanning (Deep iScan), which combines iPSC and AI, will contribute to supporting the diagnosis of ALS, and its use for other diseases may be considered in the future.

To date, we have also been working on the development of a drug discovery platform using disease-specific phenotypes of iPSC and AI. The process of drug discovery is long, and the research and development of a single new drug can take more than 10 years. Various attempts are being made to make this process more

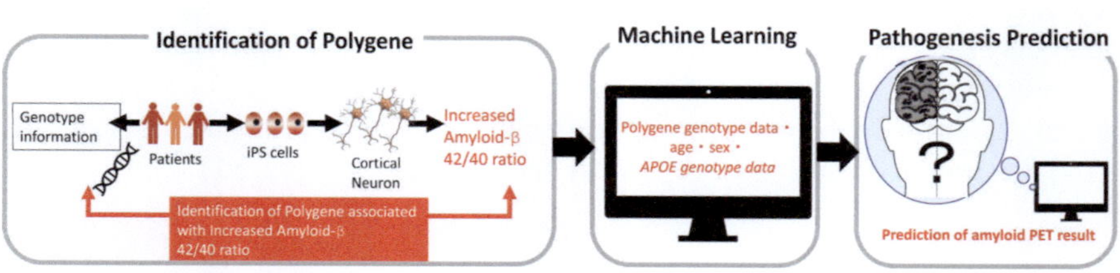

Fig. 10.2 Alzheimer's disease pathogenesis prediction via CDiP

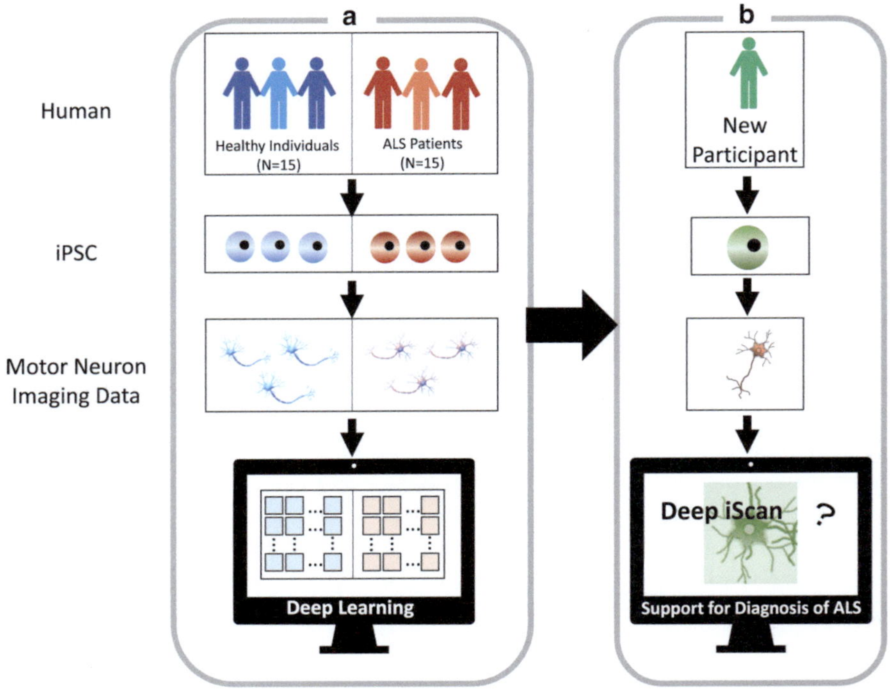

Fig. 10.3 Identification/diagnosis of ALS via Deep iScan

efficient and accelerated, including the use of AI. We would like to introduce our research aimed at streamlining the compound screening process using iPSC and AI.

Compound screening is an essential process for identifying therapeutic candidates, but evaluating millions of compounds requires significant effort and expense. Therefore, we developed a machine learning algorithm named prediction model based on the Heat-Diffusion Equation (PM-HDE) and reported that it is possible to perform efficient large-scale compound screening [34]. Specifically, we screened approximately 50,000 compounds for compounds that suppress cell death in motor neurons generated from ALS patient iPSC and built an AI that learned from the results. Using this AI, we conducted a virtual screening to predict the activity of approximately two million compounds. When we investigated the actual motor-neuron death in a dish, suppressing activities of the compounds predicted to have activities against motor neurons, it became clear that those compounds could be efficiently selected (Fig. 10.4). These results demonstrate that this model is effective in predicting hit compounds and are an example of how the new drug development process can actually be made more efficient by applying new technologies such as AI to the drug discovery process. Further applications are expected [32].

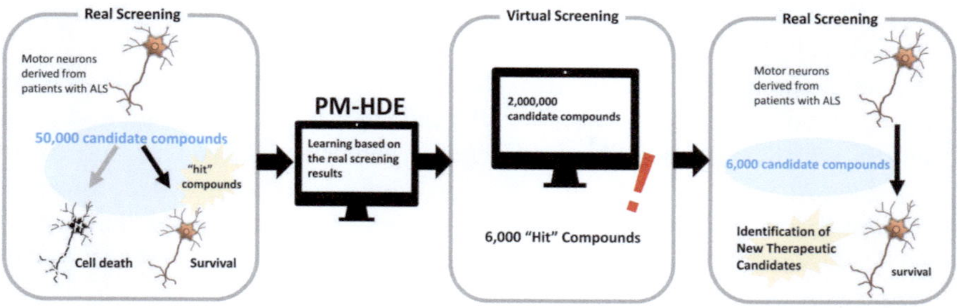

Fig. 10.4 Virtual screening of ALS therapeutic compound candidates using PM-HDE model

10.5 Conclusion

Aging is the greatest risk factor for neurodegenerative diseases. As numerous countries such as Japan advance toward a super-aging societies, together with the advancement in the diagnostic technology, the number of patients with neurodegenerative diseases will continue to increase. Therefore, early diagnosis and development of therapeutic methods for neurodegenerative diseases are among the most urgent issues. In this paper, we presented the possibility that research using iPSC data and AI will lead to the development of methods to support the early diagnosis of neurodegenerative diseases and the acceleration of efficiency in the drug discovery processes. It is hoped that iPSC and AI, two technological innovations born in different fields, will transcend the boundaries of field and contribute to the prevention, treatment, and drug discovery of diseases and ultimately to the realization of a disease-free society.

Acknowledgments We would like to thank Dr. Naonori Ueda for his guidance in AI research. We would like to thank Drs. Takayuki Kondo, Takeshi Ikeuchi, Norikazu Hara, Yuichiro Yada, Satoshi Koyama, Ryo Yamada, Takashi Asada, Tetsuaki Arai, and Kenji Ishii or their contribution to developing CDiP technology. We would also like to express our deepest gratitude to the J-ANDI research (research representative: Dr. Takeshi Iwatsubo) and ADNI research (research representative: Dr. Michael W. Weiner) for providing valuable data. We would like to thank Drs. Yuichiro Yada, Yuishin Izumi, Mitsuya Morita, Akihiro Kawada, Takayo Arisato, Hidefumi Kawakami, Masanori Nakagawa, and Ryosuke Takahashi for their contribution in developing Deep iScan. Thank you. We would like to thank Drs. Tadashi Hidaka, Takeshi Hioki, Terufumi Takagi, Yoshikazu Giga, Miho Giga, Yoshinobu Kawahara, and Makoto Fushimi for their contribution to the development of PM-HDE. These studies were supported by AMED under Grant Numbers JP22bm0804034, JP23bm1423014, JP23bm1423012, JP23bm1223013, JP24wm0625501, JP23bm1323001, JP23dk0207066, and JP24bm1123047.

Glossary

Alzheimer's Disease Neuroimaging Initiative (ADNI) ADNI and J-ADNI operate in the United States and Japan, respectively, with the aim of early diagnosis, understanding the progression mechanism, and development of treatments for

AD. This study used specific molecules in AD to track pathological changes in AD and collect related information.

Study participants included healthy older adults with normal cognitive function (control group), older adults with mild cognitive impairment, and older adults with AD, who underwent regular cognitive assessment and brain imaging. As tests and biological samples continue to be provided, information on the onset and progression of AD has been accumulated. These data are made available to researchers and medical professionals and are used to research AD and develop treatments.

CDiP An advanced analytical method to understand the cellular mechanisms underlying complex diseases under influence of multiple genetic variants (polygenicity).

HDE A deep learning-based framework to learn multiple hierarchical nonlinear kinetic slow mode.

LightGBM A high-performance, distributed, and efficient gradient boosting framework that is primarily used for classification, regression, and ranking tasks. It was developed by Microsoft and is designed to be fast and scalable, making it particularly well-suited for large datasets and high-dimensional data.

Polygene A group of genes that constitute the genetic factors for a trait that cannot be explained by Mendel's laws. The influence of individual genetic factors is small, but the effects of many of these factors add up and appear as a trait.

Random Forest An ensemble machine learning algorithm used for classification, regression, and other tasks that operate by constructing multiple decision trees during training and outputting the mode of the classes (for classification) or the mean prediction (for regression) of the individual trees.

Single cell proteomics (SCP) A cutting-edge field of research focused on studying the complete set of proteins expressed in individual cells; this approach provides insights into cellular heterogeneity and the specific roles proteins play in various biological processes at the single-cell level.

Single nucleotide polymorphisms (SNPs) SNPs are the most common type of genetic variation among people. A SNP represents a difference in a single nucleotide—A, T, C, or G—in the DNA sequence. These variations occur approximately once in every 300 nucleotides, which means there are roughly 10 million SNPs in the human genome.

UMAP Used to visualize high-dimensional data in the fields such as bioinformatics, machine learning, and data science for visualizing complex datasets and identifying patterns or clusters within the data.

References

1. Takahashi K, Yamanaka S. Induction of pluripotent stem cells from mouse embryonic and adult fibroblast cultures by defined factors. Cell. 2006;126(4):663–76. https://doi.org/10.1016/j.cell.2006.07.024.

2. Takahashi K, Tanabe K, Ohnuki M, et al. Induction of pluripotent stem cells from adult human fibroblasts by defined factors. Cell. 2007;131(5):861–72. https://doi.org/10.1016/j.cell.2007.11.019.
3. Inoue H, Yamanaka S. The use of induced pluripotent stem cells in drug development. Clin Pharmacol Ther. 2011;89(5):655–61. https://doi.org/10.1038/clpt.2011.38.
4. Inoue H, Nagata N, Kurokawa H, Yamanaka S. iPS cells: a game changer for future medicine. EMBO J. 2014;33(5):409–17. https://doi.org/10.1002/embj.201387098.
5. Yamanaka S, Inoue H, Wu J, et al. Induced pluripotent stem cell technology: a decade of progress. Nat Rev Drug Discov. 2017;16:115–30. https://doi.org/10.1038/nrd.2016.245.
6. Kondo T, Funayama M, Miyake M, et al. Modeling Alexander disease with patient iPSCs reveals cellular and molecular pathology of astrocytes [published correction appears in Acta Neuropathol Commun 2016 Sep 16;4(1):101. doi: 10.1186/s40478-016-0366-8]. Acta Neuropathol Commun. 2016;4(1):69. Published 2016 Jul 11. https://doi.org/10.1186/s40478-016-0337-0.
7. Murakami N, Imamura K, Izumi Y, et al. Proteasome impairment in neural cells derived from HMSN-P patient iPSCs. Mol Brain. 2017;10(1):7. Published 2017 Feb 15. https://doi.org/10.1186/s13041-017-0286-y.
8. Hinton GE, Osindero S, Teh YW. A fast learning algorithm for deep belief nets. Neural Comput. 2006;18(7):1527–54. https://doi.org/10.1162/neco.2006.18.7.1527.
9. Israel MA, et al. Probing sporadic and familial Alzheimer's disease using induced pluripotent stem cells. Nature. 2012;482:216–20.
10. Kondo T, et al. Modeling Alzheimer's disease with iPSCs reveals stress phenotypes associated with intracellular aβ and differential drug responsiveness. Cell Stem Cell. 2013;12:487.
11. Muratore CR, et al. The familial Alzheimer's disease APPV717I mutation alters APP processing and Tau expression in iPSC-derived neurons. Hum Mol Genet. 2014;23:3523–36.
12. Sánchez-Danés A, et al. Disease-specific phenotypes in dopamine neurons from human iPS-based models of genetic and sporadic Parkinson's disease. EMBO Mol Med. 2012;4:380–95.
13. Devine MJ, et al. Parkinson's disease induced pluripotent stem cells with triplication of the α-synuclein locus. Nat Commun. 2011;2:440.
14. Burkhardt MF, et al. A cellular model for sporadic ALS using patient-derived induced pluripotent stem cells. Mol Cell Neurosci. 2013;56:355–64.
15. Imamura K, et al. The Src/c-Abl pathway is a potential therapeutic target in amyotrophic lateral sclerosis. Sci Transl Med. 2017;9:eaaf3962.
16. Guo W, et al. HDAC6 inhibition reverses axonal transport defects in motor neurons derived from FUS-ALS patients. Nat Commun. 2017;8:861.
17. Oksanen M, et al. PSEN1 mutant iPSC-derived model reveals severe astrocyte pathology in Alzheimer's disease. Stem Cell Rep. 2017;9:1885–97.
18. Domenico A, et al. Patient-specific iPSC-derived astrocytes contribute to non-cell-autonomous neurodegeneration in Parkinson's disease. Stem Cell Rep. 2019;12:213–29.
19. Ferraiuolo L, et al. Oligodendrocytes contribute to mumotor neuron death in ALS via SOD1-dependent mechanism. Proc Natl Acad Sci. 2016;113:E6496–505.
20. Nonaka H, Kondo T, Suga M, et al. Induced pluripotent stem cell-based assays recapture multiple properties of human astrocytes. J Cell Mol Med. 2024;28:e18214. https://doi.org/10.1111/jcmm.18214.
21. Stamate D, et al. A metabolite-based machine learning approach to diagnose Alzheimer-type dementia in blood: results from the European medical information framework for Alzheimer disease biomarker discovery cohort. Alzheimers Dement Transl Res Clin Interv. 2019;5:933–8.
22. Giorgio J, et al. Modelling prognostic trajectories of cognitive decline due to Alzheimer's disease. Neuroimage Clin. 2020;26:102199.
23. Yang Y, et al. Artificial intelligence-enabled detection and assessment of Parkinson's disease using nocturnal breathing signals. Nat Med. 2022;28:2207–15.
24. Yin B, et al. Using the structure of genome data in the design of deep neural networks for predicting amyotrophic lateral sclerosis from genotype. Bioinformatics. 2019;35:i538–47.

25. Zhang S, et al. Genome-wide identification of the genetic basis of amyotrophic lateral sclerosis. Neuron. 2022;110:992–1008.e11.
26. Watanabe H, et al. A rapid functional decline type of amyotrophic lateral sclerosis is linked to low expression of TTN. J Neurol Neurosurg Psychiatry. 2016;87:851.
27. Faghri F, et al. Identifying and predicting amyotrophic lateral sclerosis clinical subgroups: a population-based machine-learning study. Lancet Digit Health. 2022;4:e359–69.
28. Coulombe B, Durcan TM, Bernard G, et al. The 37TrillionCells initiative for improving global healthcare via cell-based interception and precision medicine: focus on neurodegenerative diseases. Mol Brain. 2024;17:18. https://doi.org/10.1186/s13041-024-01088-4.
29. Kondo T, Yada Y, Ikeuchi T, Inoue H. CDiP technology for reverse engineering of sporadic Alzheimer's disease. J Hum Genet. 2023;68:231–5.
30. Pancotti C, Birolo G, Rollo C, et al. Deep learning methods to predict amyotrophic lateral sclerosis disease progression. Sci Rep. 2022;12:13738. https://doi.org/10.1038/s41598-022-17805-9.
31. Schiff L, Migliori B, Chen Y, et al. Integrating deep learning and unbiased automated high-content screening to identify complex disease signatures in human fibroblasts. Nat Commun. 2022;13:1590. https://doi.org/10.1038/s41467-022-28423-4.
32. Kondo T, et al. Dissection of the polygenic architecture of neuronal Aβ production using a large sample of individual iPSC lines derived from Alzheimer's disease patients. Nat Aging. 2022;2:125–39.
33. Imamura K, et al. Prediction model of amyotrophic lateral sclerosis by deep learning with patient induced pluripotent stem cells. Ann Neurol. 2021;89:1226–33.
34. Hidaka T, et al. Prediction of compound bioactivities using heat-diffusion equation. Patterns. 2020;1:100140.

MIX
Papier aus verantwortungsvollen Quellen
Paper from responsible sources
FSC® C105338

If you have any concerns about our products,
you can contact us on
ProductSafety@springernature.com

In case Publisher is established outside the EU,
the EU authorized representative is:
Springer Nature Customer Service Center GmbH
Europaplatz 3, 69115 Heidelberg, Germany

Printed by Libri Plureos GmbH
in Hamburg, Germany